Advances in Anatomy
Embryology and Cell Biology

Vol. 76

W0050434

Editors
F. Beck, Leicester W. Hild, Galveston
J. van Limborgh, Amsterdam R. Ortmann, Köln
J.E. Pauly, Little Rock T.H. Schiebler, Würzburg

Peter Kugler

On Angiotensin-Degrading Aminopeptidases in the Rat Kidney

With 88 Figures

Springer-Verlag
Berlin Heidelberg New York 1982

Priv.-Doz. Dr. med. Peter Kugler
Institute of Anatomy
University of Würzburg
Koellikerstraße 6
D-8700 Würzburg
Federal Republic of Germany

Supported by the Deutsche Forschungsgemeinschaft (SFB 105)

ISBN-13: 978-3-540-11452-9   e-ISBN-13: 978-3-642-68552-1
DOI: 10.1007/978-3-642-68552-1

Library of Congress Cataloging in Publication Data
Kugler, Peter, 1948 – On angiotensin-degrading aminopeptidases in the rat
kidney. (Advances in anatomy, embryology, and cell biology; v. 76) Biblio-
graphy: p. Includes index. 1. Aminopeptidases-Analysis. 2. Angiotensin-
Metabolism. 3. Cytochemistry. 4. Kidneys. 5. Rats-Physiology. I. Title.
II. Series. [DNLM: 1. Aminopeptidases-Metabolism. 2. Kidney-Physiology.
3. Rats-Physiology. W1 AD433K v. 76 / QU 136 K950]
QP609.A44K84   599.01'9256   82-5523
ISBN-13: 978-3-540-11452-9   AACR2

© Springer-Verlag Berlin Heidelberg 1982

Composition: Schreibsatz-Service Weihrauch, Würzburg

2121/3321-543210

Dedicated in gratitude to Prof. Dr. T.H. Schiebler

Contents

1 Introduction

The octapeptide angiotensin II (ANG II, Fig. 1) is the key effector substance of the renin-angiotensin system (RAS) (Werning 1972, Page and Bumpus 1974, Hierholzer 1977, Vecsei et al. 1978, Johnson and Anderson 1980 lit.). ANG II is formed in two enzymatic steps. Renin acts on renin substrate, a glycoprotein, to produce angiotensin I (ANG I, a decapeptide), which in turn is acted upon by converting enzyme to form ANG II (Skeggs et al. 1968, Fig. 1).

Renin substrate (angiotensinogen) is produced mainly in the liver (Page et al. 1941) and is a constituent of the α-globulin fraction in the circulating plasma (Plentl et al. 1943). The two enzymes involved in the formation of ANG II from renin substrate are formed at various sites in the body. Renin (E.C. 3.4.99.19) is produced mainly in the granular epithelioid cells of the kidney (Cook 1971, Taugner et al. 1979, Davidoff and Schiebler 1981), and converting enzyme (CE, E.C. 3.4.15.1) occurs chiefly in the lung (Ng and Vane 1967, Bakhle 1974 lit.) as well as in numerous other tissues, such as the juxtaglomerular apparatus of the kidney (Granger et al. 1969, 1972) and the brush border of the renal proximal tubule (Ward et al. 1975, 1976; Ward und Erdös 1977).

The biological effects of ANG II are numerous. Its best known effects are vascular (vasoconstriction), adrenal (release of catecholamines and the mineralocorticoud aldosterone), and renal (decrease in renal blood flow and glomerular filtration rate, increase in filtration fraction, varying effects on sodium and water excretion and tubulo-glomerular balance) (Page and Bumpus 1974, Hierholzer 1977 lit.).

Angiotensin II has a very short half-life equal to about one circulation time (Hodge et al. 1967). According to previous studies, this means a rapid deactivation of the peptide hormone by enzymatic degradation (Ryan 1974 lit.). The enzymes that degrade the angiotensins are called angiotensinases (E.C. 3.4.99.3; Figs. 1, 2), a term not

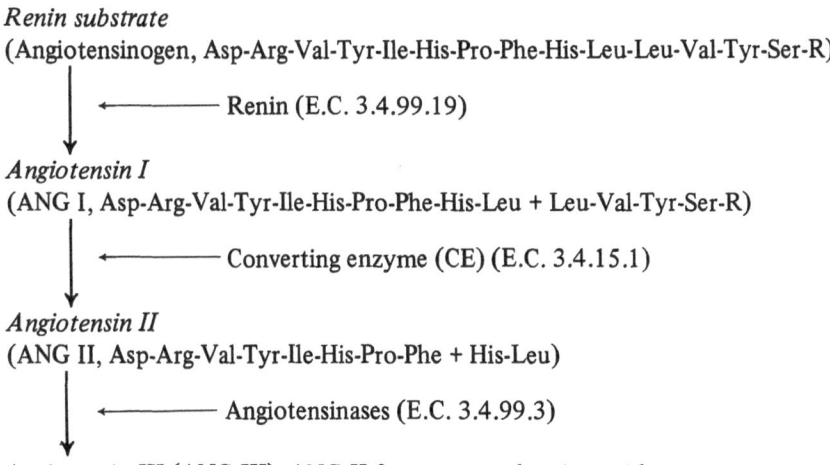

Renin substrate
(Angiotensinogen, Asp-Arg-Val-Tyr-Ile-His-Pro-Phe-His-Leu-Leu-Val-Tyr-Ser-R)

  ⟵———— Renin (E.C. 3.4.99.19)

Angiotensin I
(ANG I, Asp-Arg-Val-Tyr-Ile-His-Pro-Phe-His-Leu + Leu-Val-Tyr-Ser-R)

  ⟵———— Converting enzyme (CE) (E.C. 3.4.15.1)

Angiotensin II
(ANG II, Asp-Arg-Val-Tyr-Ile-His-Pro-Phe + His-Leu)

  ⟵———— Angiotensinases (E.C. 3.4.99.3)

Angiotensin III (ANG III), ANG II fragments and amino acids

Fig. 1. Biochemical relationships in the renin-angiotensin system (modified from Skeggs et al. 1968)

intended to denote specificity (Ledingham and Leary 1974). The products of ANG II degradation are angiotensin III (ANG III) as well as ANG II fragments and amino acids (Fig. 1).

While angiotensinase activities are demonstrable in both the blood and tissues (Ryan 1974 lit.), angiotensin degradation takes place mainly in the various body tissues (Hodge et al. 1967). One of the most important of these sites is the kidney (Hodge et al. 1967, Leary and Ledingham 1969, Carone et al. 1980). Thus, the kidney is both a target organ and degradation site for ANG II. It has been shown that the effects of the peptide hormones are regulated less by their synthesis than by their degradation (Carone et al. 1980). Thus, in studying the regulation of ANG II effects in the kidney, it is important to know which enzymes degrade angiotensins in the kidney and where they are localized. It is known from biochemical studies that the following enzymes are available in the kidney for the degradation of angiotensins (Fig. 2): (a) Aminopeptidase A (Glenner et al. 1962), called also angiotensinase A_2 (Nagatsu et al. 1965, 1970; Khairallah and Page 1967), which splits off the N-terminal amino acid L-aspartic acid from ANG II; (b) carboxypeptidase A or angiotensinase C, which removes the C-terminal amino acid L-phenylalanine from ANG II (Yang et al. 1968, Matsunaga 1971); (c) an endopeptidase, angiotensinase B, which splits the Tyr-Ile peptide bond in ANG II (Regoli et al. 1963, Matsunaga 1971), forming two tetrapeptides; and (d) unidentified aminopeptidases which split off N-terminal amino acids from

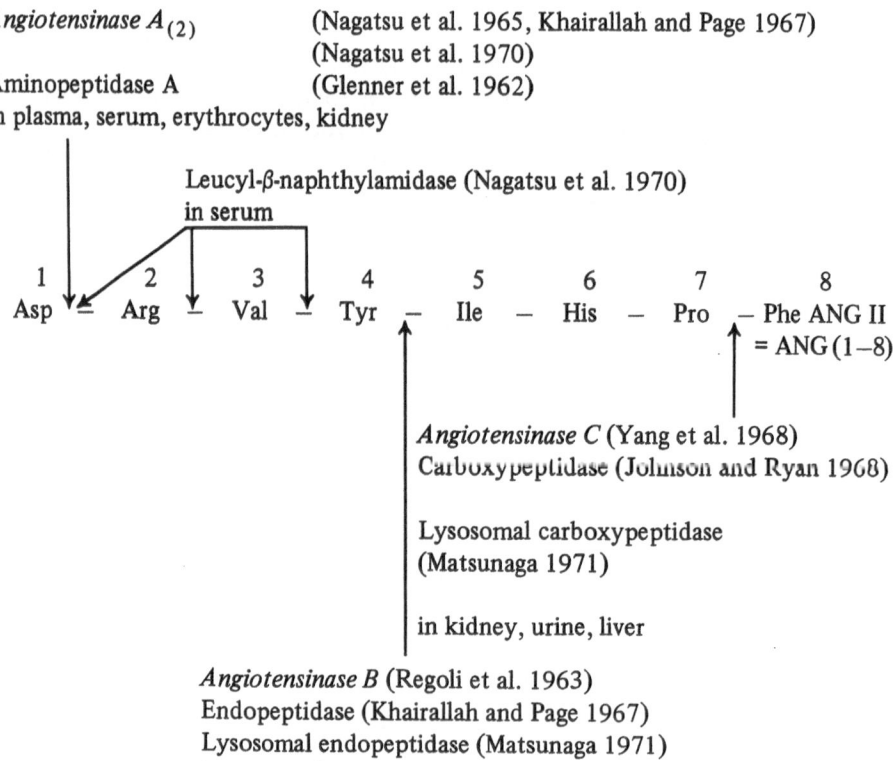

Angiotensinase $A_{(2)}$ (Nagatsu et al. 1965, Khairallah and Page 1967)
                                          (Nagatsu et al. 1970)
Aminopeptidase A   (Glenner et al. 1962)
in plasma, serum, erythrocytes, kidney

Leucyl-β-naphthylamidase (Nagatsu et al. 1970)
in serum

Angiotensinase C (Yang et al. 1968)
Carboxypeptidase (Johnson and Ryan 1968)

Lysosomal carboxypeptidase
(Matsunaga 1971)

in kidney, urine, liver

Angiotensinase B (Regoli et al. 1963)
Endopeptidase (Khairallah and Page 1967)
Lysosomal endopeptidase (Matsunaga 1971)
in serum, plasma, kidney, liver

Fig. 2. Enzymatic degradation of angiotensin II. The *arrows* indicate the sites of enzyme action

ANG II (Hess 1965, Matsunaga and Masson 1970). Osborne et al. (1970) have shown in perfusion studies of isolated rat kidneys that ANG II is degraded in the kidney chiefly by aminopeptidases and endopeptidases, because the renal perfusates mainly contained 2–8 heptapeptide, 3–8 hexapeptide and 1–4 tetrapeptide (cf. Fig. 2).

With regard to the cellular or subcellular localization of these angiotensinases, biochemical fractionations to date have shown only that they occur in the mitochondria and microsomes (Dengler and Reichel 1960) or in the lysosomes and microsomes (Matsunaga et al. 1968, 1969; Saito et al. 1969). So far the only one of the foregoing peptidases that has been histochemically localized is aminopeptidase A (APA, E.C. 3.4.11.7; Glenner and Folk 1961, Glenner et al. 1962, Lojda and Gossrau 1980), which splits α-L-glutamic acid-2-naphthylamide (α-L-Glu-2NA). This membrane-bound enzyme was found by light microscopy in the glomerulus and brush borders of the proximal tubule (Glenner et al. 1962, Lojda and Gossrau 1980). While it could be shown biochemically that the α-L-Glu-2NA-splitting enzyme is an angiotensinase (Hess 1965), it could not be proved that the histochemically localized APA functions as an angiotensinase. Another renal aminopeptidase that might be important as an angiotensinase is aminopeptidase M (APM, E.C. 3.4.11.2; cf. Delange and Smith 1971, Lojda et al. 1979), which, like APA, is demonstrable in the brush borders of the proximal tubule but is not found in the glomerulus (Wachsmuth 1968, Wachsmuth and Torhorst 1974, Wachsmuth and Donner 1976). APM, which is also membrane-bound, reportedly splits all peptides with a free amino group and L-amino acids, with the preferential removal of L-alanine (Delange and Smith 1971 lit; cf. Lojda et al. 1979).

Because aminopeptidases play an essential role in the breakdown of angiotensins and thus in the regulation of angiotensin effects in the kidney, we have investigated APA and APM in the rat and mouse kidney by biochemical and various histochemical methods. Rat kidneys were mainly used in our methodological studies. Our objectives were:

1. To determine the optimum conditions for the histochemical demonstration of APA

2. To differentiate between APA and APM in the kidney on the basis of their different ion dependences by means of biochemical and histochemical methodological studies

3. To determine the influence of ANG I, II, and III on APA by quantitative histochemical measurements in the glomerulus in order to realize the significance of the histochemically demonstrable APA in angiotensin degradation; the glomerulus was selected as the measuring site, since it contains no APM that could interact with the APA

4. To determine the types of enzyme inhibition produced by ANG I, II, and III, using fluorometric measurements in renal homogenate for APA and APM, and in microdissected glomeruli for APA; these measurements in the homogenate and glomeruli (cf. 3) allow a comparison of reaction sites in the kidney where APA and APM occur together (homogenate) or APA occurs alone (glomeruli)

5. To study the localization of APA in the kidney by light microscopy and ultracytochemistry, perhaps discovering previously unrecognized reaction sites

6. To perform histochemical and biochemical APA and APM determinations in the kidneys of experimental animals (male rats and mice) in order to clarify the regulation of APA and APM and its importance for the renin-angiotensin system in the kidney

We selected only those experimental conditions that are associated with known, clear-cut effects in the RAS. These include adrenalectomy (Gross and Sulzer 1957, Schaechtelin et al. 1963, Gross et al. 1965, Danda and Deveny 1971, Bucher and Kaissling 1973 lit.) and a low-sodium diet (Bareiss and Kracht 1969; Bucher and Kaissling 1973 lit, Nishimura 1980 lit, Nakane et al. 1980), which lead to increased renin production, as well as a high-sodium diet, which decreases renin production (Bareiss and Kracht 1969, Bucher and Kaissling 1973 lit; Nishimura 1980 lit; Nakane et al. 1980). Our goal was to explore the regulation of APA and APM and its importance for the RAS in the kidney.

2 Materials and Methods

2.1 Material

The investigations were performed on 350 Wistar rats (330 males, 20 females) and 100 NMRI mice (75 males, 25 females) of our own breeding (random-bred closed colony). The animals were housed in Makrolon cages (four rats per cage, four to six mice per cage) at $21° \pm 2 °C$ with a 12-h light-dark cycle (light from 7 a.m. to 7 p.m., dark from 7 p.m. to 7 a.m.) and were given tap water and standard Altromin diet (TPF 1320, pelleted; mean dietary sodium 0.2%) ad libitum. The animals were selected according to age and weight (rats and mice 75–95 days old; male rats 280–300 g, female rats 190–210 g, male mice 35–45 g, female mice 30–40 g). Only estrous females were used, as determined by the cytologic examination of vaginal smears (Allen 1922).

Only males within the specified age and weight range were used in the animal experiments. Postoperatively or for experimental purposes, the animals were kept individually in Makrolon cages under the conditions stated above.

Animal Experiments (five to ten male rats and mice per experiment)

Low-Sodium Diet (Altromin diet C 1036: 300 mg sodium and 1000 mg chloride per kilogram diet). Diet and bidistilled water ad libitum; animals killed on day 20 of diet.

High-Sodium Diet (Altromin control diet C 1000: 0.35% mean dietary sodium). Diet and 1% sodium chloride solution (NaCl dissolved in bidistilled water) ad libitum; animals killed on day 20 of diet.

Control Animals. Control rats and mice were maintained on Altromin control diet C 1000 (0.35% mean dietary sodium) and bidistilled water ad libitum and were also killed on day 20 of the experiment.

Bilateral Adrenalectomy. Laparotomy was performed unter ether anesthesia on the dorsal side, bilaterally from the spinal column, and the adrenal glands were bluntly removed with bent forceps under the stereomicroscope. After the operation the animals were given Altromin standard diet TPF 1320 and tap water ad libitum. The mice were killed on postoperative day 7, and the rats on postoperative day 9.

Control Animals. The controls were subjected to the same operative and postoperative conditions as the experimental animals, except that the adrenal glands were not removed (sham operation).

Kidney Removal. After determination of live weight, the normal and experimental animals were (unless otherwise indicated) decapitated under light ether anesthesia between 11 a.m. and 12 noon for histochemical and biochemical study. Then the kidneys were quickly removed by laparotomy and treated as described below. The wet weight of the adrenal glands was determined for at least five normal and five experimental animals.

Only normal male animals (rats) were used for the biochemical and histochemical methodological studies unless otherwise indicated.

2.2 Qualitative Histochemistry

2.2.1 Light Microscopy

After removal of the left kidney, its cranial or caudal pole was sharply removed. The kidneys were placed on specimen holders, wrapped in transparent film, and frozen in N_2-cooled propane (Winckler 1970a). Sections $2-10$ μm thick were prepared in a cryostat (System Dittes-Duspiva) at $-25°$ to $-30°C$.

Tissue Pretreatment

Cryostat Sections. The enzyme demonstrations were performed on un-pretreated, acetone-pretreated, freeze-dried, or aldehyde-fixed cryostat section. The cryostat sections were mounted before or after pretreatment on room-temperature slides that had been cleaned for 24 h in equal parts of 99% ethanol and 100% diethyl ether. For acetone pretreatment the mounted cryostat sections were immersed in 100% analytical-grade acetone for 5 min at $-20°C$ and then air-dried briefly before the enzyme demonstration was performed. Freeze-dried cryostat sections (10 μm thick) were obtained by the Winckler method (1970b). The freeze drying was carried out for 70 min at $-50°C$ and 10^{-1} to 10^{-2} mmHg in a Leybold-Heraeus apparatus. The freeze-dried sections were mounted on albuminized slides with a 0.5% celloidin solution (FDC sections) (Lojda et al. 1979). For aldehyde fixation, mounted 10-μm cryostat sections were fixed in formol (4% freshly prepared formaldehyde with and without 1% calcium chloride in 0.1 M tris-maleate buffer, pH 7.2) or in glutaraldehyde (2% glutaraldehyde under nitrogen with and without 1% calcium in 0.1 M tris-maleate buffer, pH 7.2) for 5 min at + 4 °C. After the sections were washed for 5 min in the same buffer and briefly dried in air, the enzyme demonstration was performed.

Block Fixation. In addition, thin slices of unpretreated kidney were fixed for 20 h in formol (4% freshly prepared formaldehyde with 1% calcium chloride in 0.1 M tris-maleate buffer, pH 7.2) or in glutaraldehyde (2% glutaraldehyde with 1% calcium chloride in 0.1 M tris-maleate buffer, pH 7.2) at +4 °C. They were then washed in Holt's mixture (Holt 1959, cf. Lojda et al. 1979) for 20 h at +4 °C. The tissue slices were then frozen by the method described above, sectioned in the cryostat (10 μm), and mounted on albuminized slides (cf. Lojda et al. 1979).

Enzyme Demonstrations

All enzymes were demonstrated with the simultaneous azo coupling technique using various coupling agents (cf. Lojda et al. 1979). The reaction medium was dripped onto the sections, which were then incubated in a thermostatically controlled water bath at +30 °C. Following incubation the sections were washed for $1-3$ min in distilled water, fixed for $10-20$ min in 4% aqueous formol (this was done with all sections, since formol changes the color quality), rinsed briefly in distilled water, and mounted in Karion F or glycerin gelatin.

Aminopeptidase A (angiotensinase A; APA; E.C. 3.4.11.7)
Reaction scheme (simultaneous azo coupling):

Standard Incubation Medium (cf. Kugler 1981b). One milliliter contains: 0.5 mg α-L-Glu-MNA (final concentration 1.65 mM) dissolved in 15 μl DMF (final concentration 1.5% v/v) and 35 μl 0.4 M tris stock solution by prolonged shaking; 2.5 μl of a 1 M calcium chloride solution (final concentration 2.5 mM), 0.5 mg Fast Blue B ∘ BF$_4$ salt, purest grade (final concentration 1.1 mM; added to reaction medium shortly before incubation), 97.5 μl bidistilled water, 850 μl tris-maleate buffer, pH 6.5 (buffer preparation: bring 0.2 M tris stock to pH 6.5 with 0.2 M maleic acid anhydrite, dilute to 0.1 M tris with bidistilled water). After the components are mixed, the pH of the reaction medium is 7.

The incubation times were 5−10 for un-pretreated sections, 10−20 min for acetone-pretreated sections, 15−30 min for section-fixed tissue, 10−30 min for block-fixed tissue, and up to 60 min for FDC sections.

Determination of Optimum Fast Blue B or Fast Garnet GBC Concentration in APA Standard Incubation Medium. Varying amounts of the highly water-soluble Fast Blue B (FBB) or Fast Garnet GBC (FGGBC) were added to the APA incubation medium. Two different grades of FBB (pure and purest) and pure-grade FGGBC were used. The final concentrations ranged from 0.3 mM (0.125 mg/ml) to 3.7 mM (1.5 mg/ml) for pure-grade FBB, 0.28 mM (0.125 mg/ml) to 2.26 mM (1 mg/ml) for purest-grade FBB, and 0.82 mM (0.25 mg/ml) to 8.2 mM (2.5 mg/ml) for pure-grade FGGBC. The investigation was done on 10-μm acetone-pretreated cryostat sections using increments of 0.25 mg/ml coupling agent.

Alternative Coupling Agents. (a) Hexazotized new fuchsin (HNF; cf. Lojda et al. 1979). Of a freshly prepared 2% solution, 0.025 ml/ml incubation solution (pH 7) was used. Only un-pretreated and FDC sections were investigated. The incubation time was 15−60 min. (b) Hexazonium-*p*-rosaniline (HPR; cf. Lojda et al. 1979). Of a freshly prepared 2% solution, 0.020 ml/ml incubation solution (pH 7) was used. Only fresh and FDC sections were incubated. The incubation time was 15−60 min. In addition, APA was demonstrated at pH 5 in FDC sections with this incubation medium (HPR; incubation time up to 60 min).

Varying Ionic and Substrate Composition. Only acetone-pretreated cryostat sections were incubated. The standard medium contained:
1. Instead of α-L-Glu-MNA: 1.65 mM α-L-Glu-2NA or
α-L-Asp-MNA (in rat and mouse) or α-L-Asp-Arg-MNA
or L-Ala-MNA or L-Leu-MNA or L-Arg-MNA

2. No substrate
3. 1.5, 5, and 10 mM CaCl$_2$
4. No CaCl$_2$
5. 1.5, 5, and 10 mM EDTA
6. 1 mM 1,10-phenanthrolene

Peptides. Addition of substrate-equimolar amounts of ANG I or ANG II or ANG III or L-Tyr-Ile or L-His-Pro-Phe to APA incubation medium (all peptides, with the exception of L-Tyr-Ile, are highly water-soluble over the concentration range indicated). Only acetone-pretreated cryostat sections were investigated.

Buffers. Instead of tris-maleate buffer, 0.1 M cacodylate buffer, pH 6.5 (pH of complete reaction medium = 7) or 0.1 M tris-HCl pH 7 was also used. Only acetone-pretreated cryostat sections were incubated.

Aminopeptidase M (APM; E.C. 3.4.11.2)
Reaction scheme (simultaneous azo coupling):

Standard Medium (cf. Lojda et al. 1979). One milliliter contains: 0.37 mg L-Ala-MNA (final concentration 1.5 mM) dissolved in 18 μl DMF (final concentration 1.8% v/v), 982 μl tris-maleate buffer, pH 6.5 (buffer preparation see APA), and 0.5 mg purest-grade FBB or 1 mg pure-grade FBB (added shortly before start of reaction).
Incubation times: 5–10 min for un-pretreated and acetone-pretreated sections, 15–30 min for block-fixed tissue, 15–30 min for aldehyde-fixed sections, and 30–60 min for FDC sections.
The following investigations were done only on acetone-pretreated cryostat sections:

Varying Ionic and Substrate Composition. The standard medium contained:
1. Instead of L-Ala: L-Leu-MNA or L-Arg-MNA or L-Pro-MNA (incubation solution flocculates, filtration) or α-L-Asp-Arg-MNA (with and without 2.5 mM CaCl$_2$ in incubation medium)
2. No substrate
3. Additional 1.5, 5 and 10 mM CaCl$_2$
4. Additional 1.5, 5 and 10 mM EDTA
5. Additional 1 mM 1,10-phenanthrolene
Various peptides: See APA.

2.2.2 Ultracytochemistry of Aminopeptidase A

Only normal and adrenalectomized animals were used for these investigations. Laparotomy and thoracotomy were performed under ether anesthesia. The right atrium was opened and the animals were perfused through the left ventricle with rinsing solution 1 (see below) for 1–2 min and then with fixing solution (see below) for 1 min, followed by a 5-min post-rinse with solution 2 (see below).

Rinsing solution 1 (330–340 mosm; pH 7.3): 137 mM NaCl, 2.7 mM KCl, 10 mM Na$_2$HPO$_4$, 1.5 mM KH$_2$PO$_4$, 0.5 mM MgCl$_2$, 1 g/liter D-glucose, 10 g/liter saccharose; 5 g/liter procaine HCl; 0.5 ml/liter Thrombophob (5000 U.S.P. units sodium heparin per ml).

Rinsing solution 2 (300–310 mosm; pH 7.3): 0.1 M cacodylate buffer, pH 7.4 and 50 g/liter saccharose.

Fixing solution (330–340 mosm; pH 7.3): 0.5% glutaraldehyde, 0.75% freshly prepared formaldehyde, 10 mM CaCl$_2$, 0.05 M cacodylate buffer, pH 7.3.
    Following perfusion, the kidneys were cut into small pieces and collected in rinsing solution 2. These pieces were embedded in 4% agarose type VII from Sigma (low gelling temperature; dissolved in 0.1 M cacodylate buffer, pH 7) at 37 °C. After solidification of the agarose at +4 °C, 40-μm-thick slices were cut with a tissue chopper (Sorvall TC-2 Tissue Sectioner) and rinsed for 15 min in solution 2.
    The particles were then incubated on a shaker for 2 h at room temperature (incubation medium, see below) for the demonstration of APA. For control purposes, tissue sections were incubated in substrate-free medium under otherwise identical reaction conditions.

APA-Incubation Medium (pH 5, 6.5 and pH 7). Ten milliliters contain: 7.5 mg αL-Glu-MNA (final concentration 2.4 mM) dissolved in 150 μl DMF and 350 μl of 0.4 M tris, 9.15 ml of 0.1 M cacodylate buffer, pH 7, 0.1 ml of a 1 M CaCl$_2$ solution (final concentration 10 mM), and 0.25 ml of a 2% HPR or HNF solution (cf. Lojda et al. 1979), pH adjusted to 5, 6.5 (HPR) or 7 (HNF). Filtration is necessary for the pH 6.5 and pH 7 incubation mediums. Controls: incubation medium same as above but without α-L-Glu-MNA.
    After incubation, tissue slices were postfixed for 15 min at room temperature (2.5% glutaraldehyde and 2.5% freshly prepared formaldehyde in 0.05 M cacodylate buffer, pH 7.2), rinsed in 0.1 M cacodylate buffer, pH 7.3, and osmified for 14–16 h at room temperature (2% aqueous osmium tetroxide). They were then rinsed with 0.1 M cacodylate buffer, pH 7.2, dehydrated in graded acetone, and embedded in Durcupan. Ultrathin sections were cut on a Reichert OMU-3 ultramicrotome and examined with a Zeiss 9A electron microscope without prior staining with uranyl acetate or lead citrate.

2.3 Quantitative Enzyme Histochemistry of Aminopeptidase A

To obtain quantitative data on enzyme kinetics in tissue sections, we employed an instrumental setup consisting of a Vickers M 85 microdensitometer and Kontron computer-assisted Videoplan System (Kugler 1981a). The components are linked via a BCD interface. In this method the reaction products formed by enzymatic activity in a specified tissue area are measure microdensitometrically over time (enzyme kinetics), after which the reaction area is morphometrically analyzed. This makes it possible to relate measured enzyme activities to the associated reaction volume (for general procedure cf. Kugler 1981a). The methodological studies described below were performed only in male rats. Only glomeruli were used for the enzyme-kinetic measurements, as these structures are devoid of APM (cf. Wachsmuth 1968, Wachsmuth and Torhorst 1974, Kugler 1981b).

2.3.1 Determination of the Absorption Spectrum of the Azo Dye from 4-Methoxy-2-naphthylamine and Purest-Grade Fast Blue B

For this purpose, 10-μm acetone-pretreated sections were incubated in APA standard medium for 10 min at +30 °C (see above), then washed for 10 min in several changes of distilled water at

+4 °C and mounted in glycerin gelatin. Postfixing in formol was avoided, as this changes the color quality. The M85 was then used to measure weak and strong reaction sites in the kidney between 470 and 600 nm. The following microdensitometer settings were used: bandwidth 15, spot size 2, scanning frame size 1, objective 40×, gating mask A1.

2.3.2 Enzyme-Kinetic Studies

Reaction Medium (Fig. 3). The reaction medium for the APA demonstration was prepared from five different stock solutions: (a) a stock substrate solution (5 mg α-L-Glu-MNA dissolved by shaking in 150 μl DMF and 350 μl 0.4 M tris stock), (b) an equalizing solution (300 μl DMF and 700 μl 0.4 M tris stock), (c) a 1 M CaCl₂ solution, (d) an FBB stock solution (5 mg purest-grade FBB in 1 ml bidistilled water) and (e) a tris-maleate buffer pH 6.5 (preparation, see above).

Aliquots were taken from each of stock solutions a, b, c, and e, and a mixture of 900 μl was prepared. Then, shortly before incubation of the sections, 90 μl of this solution was mixed with 10 μl of FBB stock solution 4, so that 100 μl of reaction solution was available for each incubation. The incubation components were separated for reasons of medium stability (spontaneous azo formation). Alone, all five stock solutions will keep for at least 2–3 days at +4 °C.

Various aliquots were taken from the stock solutions and mixed according to the requirements of the investigation. For example, the following aliquots were mixed for an incubation

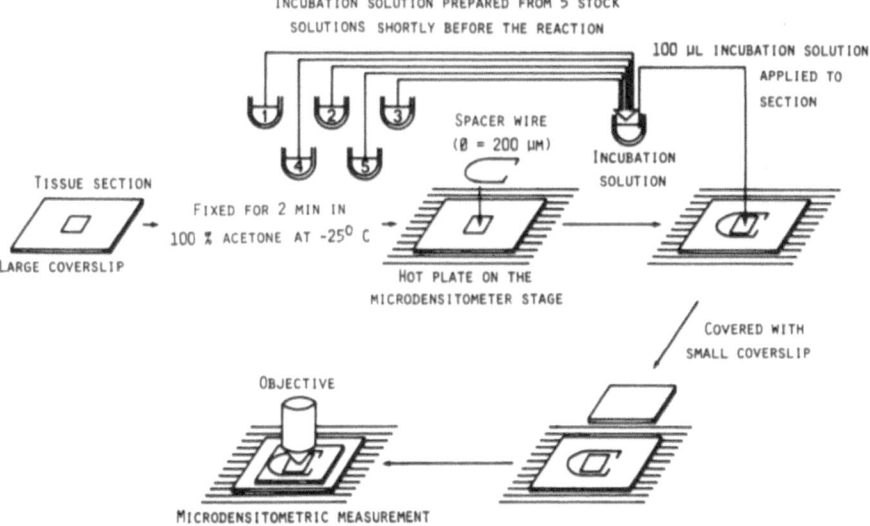

Fig. 3. Specimen preparation for enzyme-kinetic measurements in tissue sections. The renal section is mounted on a large coverslip at room temperature, cooled to −25 °C on the microtome holder (cryostat), and then fixed in 100% analytical-grade acetone for 2 min at −25 °C. This preparation is placed onto the hot plate (+30 °C) on the microdensitometer stage (drop of immersion oil between hot plate and coverslip to improve contact). At 400× magnification a subcapsular glomerulus is located in the section, and a circular measuring field (about 640 μm²) is defined on a suitable glomerular area by the gating mask. After a spacer wire (φ = 200 μm) is placed around the section on the coverslip, 100 μl of incubation solution (quantifying medium) is pipetted onto the tissue section to start the enzyme reaction. Microdensitometric measurements are performed after a small coverslip is placed onto the spacer wire to ensure an even distribution of the reaction medium on the section

9

solution with 1.5 mM α-L-Glu-MNA, 1.5 mM CaCl$_2$, and 0.5 mg/ml purest-grade FBB (quantifying medium): 45 μl stock substrate solution 1, 1.5 μl calcium chloride solution 3, 15 μl equalizing solution 2, and 838.5 μl tris-maleate buffer 5. Shortly before the start of the reaction, 90 μl of this 900-μl solution was mixed with 10 μl of FBB solution. 4. The final concentration of DMF was 1.8% in all investigations. The pH of the incubation solution is always 7 when the indicated procedure is followed.

Incubation of the Tissue Sections and Performance of the Microdensitometer Measurements (Fig. 3). Unless otherwise stated, 5-μm-thick renal sections were always used for the quantitative histochemical studies. The sections were mounted on large coverslips (60 × 24 mm; 120 μm thick) and fixed in 100% analytical-grade acetone in the cryostat for 2 min at −25 °C (to enhance localization of reaction sites). They were then placed on a special hot plate heated by flowing water, and this plate was mounted on the microdensitometer stage (Kugler 1981a). Some immersion oil was placed between the coverslip and hot plate to improve contact. All kinetic measurements were carried out at a controlled reaction temperature of +30 °C.

All measurements were performed on subcapsular glomeruli. After a glomerulus had been located with the condenser diaphragm closed, a circular gating mask about 640 μm^2 in size was projected onto it, covering a portion of the vascular loops. Next a U-shaped spacer wire about 200 μm thick was placed around the tissue section, 100 μl of freshly mixed reaction solution was pipetted onto the section, and a small coverslip (15 × 15 mm; 120 μm thick) was laid onto the spacer wire over the section. The small coverslip and spacer wire ensured a well-defined layer thickness. Now the condenser diaphragm was quickly opened, the focus was adjusted, and the computer-controlled measurement was started with a magnification of 40 at 500 nm. The flying-spot system of the M85 furnishes integrated data over 6-s measuring cycles. After the measurements were completed, a correlation analysis between density values and reaction time by the Videoplan computer yielded the slope of the reaction line as an expression of enzyme activity. Then the microscopic image of the specimen was transferred by TV camera to a monitor, and the densitometrized reaction area within the glomerulus was morphometrized in video dialog with the graphic tablet. This made it possible to relate enzyme activities to the associated reaction area (cf. Kugler 1981a).

The following parameters that are important for quantitative histochemical investigations of APA were determined by this technique:

APA Activity as a Function of Substrate Concentration (determination of K$_m$; rat and mouse glomeruli). The APA activity was determined in 5-μm-thick cryostat sections, which were incubated with the APA quantifying medium containing 0.1−1.5 mM α-L-Glu-MNA. K$_m$ and V$_{max}$ were determined by means of the Lineweaver-Burk plot (1934).

APA Activity as a Function of Time (rat glomeruli). Using the quantifying medium, APA activity was kinetically measured in 5-μm sections over a 10-min period. The period of linear APA kinetics (linear regression) was determined with the correlation analysis program of the Videoplan computer.

Influence of Section Thickness on APA Activity (rat glomeruli). Cryostat sections 3, 5, 7, 9, and 12 μm thick and the quantifying medium were used. By correlation analysis, the Videoplan computer determined the relationship between enzyme activity and section thickness (linear regression).

Influence of Acetone Pretreatment of Cryostat Sections on APA Activity (rat glomeruli). For this purpose APA activities were measured in un-pretreated and acetone-pretreated (2 min in 100% analytical-grade acetone at −25 °C) cryostat sections 5 μm thick. The quantifying medium served as the reaction medium.

APA Activity Under Various Ionic Conditions (rat glomeruli). With the quantifying medium, the APA activity was demonstrated without CaCl$_2$ and with 1.5, 5, and 10 mM CaCl$_2$ in 5-μm cryostat sections. APA activity was also determined after adding 1.5 mM EDTA, which forms ion complexes, to the reaction medium. Finally, the effect of 50−180 mM NaCl in the quantifying medium on APA activity was measured.

APA Activity with Various Substrates (rat glomeruli). Instead of α-L-Glu-MNA, 1.5 mM α-L-Asp-MNA or 1.5 mM L-Ala-MNA was used in the quantifying medium. Measurements were performed on 5-μm cryostat sections.

APA Activity with Various Buffers (rat glomeruli). Instead of tris-maleate buffer, 0.1 M cacodylate buffer pH 6.5 was used in the quantifying medium. The reaction pH was 7.

APA Activity with Various Angiotensins used to Determine the Inhibition Type (rat and mouse glomeruli). For this investigation, specific angiotensin concentrations between 0.05 and 0.15 mM were used in the quantifying medium for various substrate concentrations (cf. K_m determination). Three different angiotensins were tested: ANG I, II, and III. The type of inhibition produced by these angiotensins was determined from the Lineweaver-Burk plot (1934), and the K_i for ANG II from the secondary Lineweaver-Burk plot (Bisswanger 1979).

Control Measurements. To determine spontaneous azo formation, kinetic measurements were performed on renal sections (glomeruli) at a reaction temperature of +30 °C with an APA incubation medium that contained no substrate, using the procedure previously described. This revealed spontaneous azo formation both in the tissue section and in the overlying incubation medium. In addition, the same kinetic measurements were performed without a tissue section, and spontaneous azo formation was measured in the incubation medium only.

2.3.3 APA Activity Pattern Along the Rat Nephron (Endpoint Measurements)

Seven-micrometer-thick cryostat sections were fixed in 100% analytical-grade acetone for 2 min at −25 °C. They were then placed onto a hot plate, the incubation medium was dripped on (APA standard medium, cf. Sect. 2.2), and the sections were incubated for 1.5 min at +30 °C. After the reaction the medium was poured off, and the sections were placed in a cuvette containing 50 ml distilled water with 1.35 ml of 1 N HCl (pH 2) for 5 min at 0 °C. The specimens were then mounted in Karion F. The APA activities along the nephron were measured against APA-free regions of the renal medulla (0 value) (see Sect. 2.3.1 for microdensitometer settings). In this way the APA activities in the glomerulus and in segments S_1, S_2, and S_3 of the proximal tubule were determined. Finally, the same technique was used to measure the APA activities in the glomerulus and S_3 segment following incubation of the sections in APA standard medium with substrate-equimolar ANG II.

2.4 Biochemistry of APA and APM

The biochemical methodological studies involved the fluorometric and photometric analysis of homogenates or glomerular preparations from the rat kidney. Fluorometric measurements were made with a Foci ratio fluorometer (Farrand, New York; primary filter 363 nm, secondary filter 450 nm). The photometric studies were done with a Zeiss PM6 spectrophotometer. The photometric and fluorometric measurements were carried out at +22° and +25 °C, respectively. Fluorometry of the renal homogenate always involved kinetic enzyme activity measurements. Only 2-naphthylamide derivatives were used for the fluorometric determination of enzyme activities. Before these measurements could be made, it was first necessary to establish the relationship between fluorescent activities and various concentrations of 2-naphthylamine in the medium (standard curve). A 2-naphthylamine stock solution was prepared for this purpose (14.3 mg dissolved in 4 ml DMF). Corresponding aliquots of this stock solution were further diluted with 0.1 M cacodylate buffer or 0.1 M tris-HCl buffer pH 7 in order to establish the fluorescent activities of 0.005 to 0.125 μmol 2-naphthylamine. The 2-naphthylamine concentrations released by enzyme activity were within this range. These standard curves were used as needed to determine specific enzyme activities.

2.4.1 Fluorometry of Renal Homogenate from Male Rats

Production of Homogenate. The rats were decapitated under light ether anesthesia and laparotomy was performed. The exposed kidneys were left attached to their vascular pedicle and were rapidly decapsulated to isolate them from surrounding tissues. They were then separated from the pedicle close to the parenchyma, and their wet weight determined. Next the kidneys were wrapped in aluminum foil and twice frozen in N_2 and thawed. Each whole kidney was homogenized with a pestle homogenizer after Potter-Elvehjem at 1500 rpm (1 ml ice-cooled bidistilled water per 200 mg kidney wet weight). The homogenate was then centrifuged for 10 min at 200 rpm (corresponds to 500 g) with a bench centrifuge. The supernatant was frozen in N_2 in Eppendorf reaction vessels and stored briefly at $-55\,°C$ until needed.

Determination of APA Activity as a Function of Substrate Concentration (determination of K_m). For this and further studies, stock solutions of the different reaction components were prepared as a source for the various reaction media. Stock substrate solution: 10.89 mg α-L-Glu-2NA dissolved in 326.7 µl DMF and 762.3 µl 0.4 M tris stock. The final concentration of DMF in the medium was always 1.6% v/v. $CaCl_2$ stock: 1 M $CaCl_2$ dissolved in bidistilled water. Equalizing solution: 1 ml of this solution contained 300 µl DMF and 700 µl of 0.4 M tris stock. The buffer was 0.1 M cacodylate buffer pH 6.5. The pH of the incubation solution automatically adjusted to 7 after the various reaction components were mixed. Twenty microliters of the renal homogenate supernatant and 980 µl of reaction solution were used for each measurement. Only in determining the K_m of APA was 0.1 M tris-HCl buffer pH 7 also used instead of cacodylate.

An incubation solution with 1.5 mM α-L-Glu-2NA and 1.5 mM calcium chloride (APA fluorometric medium), taken here as an example, contained the following aliquots from the corresponding stock solutions: for 1 ml incubation medium 40.8 µl stock substrate solution, 13.7 µl equalizing solution, 1.5 µl $CaCl_2$ stock solution, 824 µl buffer, and 100 µl bidistilled water (total = 980 µl).

For the substrate-dependent APA activity measurements, this fluorometric medium was used with varying substrate concentrations, i.e., from 0.075 to 1.5 mM α-L-Glu-2NA. K_m was determined by means of the Lineweaver-Burk plot (1934).

Determination of APM Activity as a Function of Substrate Concentration (determination of K_m). The procedure for the quantification of APM was similar to that for APA. The incubation media were prepared from the appropriate stock solutions. The stock substrate solution consisted of 34.6 mg L-Ala-2NA, which was dissolved in 1.3 ml DMF and 3.05 ml of 0.4 M tris stock. The concentration of DMF in the medium was always 1.6% v/v. All measurements were made without $CaCl_2$ in the incubation solution unless otherwise indicated. The equalizing solution and buffer were the same as for APA; only cacodylate buffer was used. For 1 ml of incubation solution containing 1.5 mM L-Ala-2NA (APM fluorometric medium), the following components were mixed: 40.8 µl stock substrate solution, 13.6 µl equalizing solution, 825.6 µl buffer, and 100 µl bidistilled water. Unless otherwise stated, all measurements were performed with 20 µl renal homogenate and 980 µl reaction solution at pH 7 (the value obtained when the reaction components were mixed).

For the substrate-dependent activity measurements, 0.033–1.5 mM L-Ala-2NA was used in the APM fluorometric medium. K_m was determined from the Lineweaver-Burk plot (1934).

Determination of APA and APM Activity Under Various Ionic Conditions. APA and APM activities were measured with the APA and APM fluorometric media, with and without 1.5, 5, and 10 mM $CaCl_2$. The activities of these enzymes were also measured with 1.5, 5, and 10 mM of the complexing agent EDTA added to the medium. In some of the tests described in this section, the alternative substrate α-L-Asp-2NA was used in the APA fluorometric medium instead of α−L-Glu-2NA. The Asp substrate is also attacked by APA, though with less activity than α-L-Glu-2NA (cf. Quantitative Enzyme Histochemistry, and Lojda and Gossrau 1980).

Determination of APA and APM Activities Using Various Angiotensins to Determine K_i. The activities of APA and APM were measured with various substrate (cf. K_m determinations) and specific angiotensin concentrations in the fluorometric media.

12

ANG I, II, or III was used in concentrations between 0.0125 and 0.05 mM; measurements were made with at least two different concentrations for each angiotensin. The type of inhibition that occurred with these angiotensins was determined from the Lineweaver-Burk plot (1934), and K_i from the secondary Lineweaver-Burk plot (Bisswanger 1979).

Determination of APA and APM Activities Using Angiotensin Fragments. In this investigation the APA and APM activities were determined with 1.5 mM L-Tyr-Ile and L-His-Pro-Phe in the given fluorometric medium. The object was to discover whether APA and APM are inhibited by these peptides and, if so, to what degree.

2.4.2 Photometry of Renal Homogenate from Male Rats

Effect of Various Concentrations of Purest-Grade FFB on APA Activity. These investigations were performed with the Zeiss PM6 spectrophotometer at a wavelength of 525 nm. The incubating medium was the APA standard medium for qualitative histochemistry, which contained various concentrations of purest-grade FBB. APA activity was measured at four different FBB concentrations: 0.25, 0.5, 0.75, and 1.0 mg/ml incubation medium. In each case 50 μl of homogenate was incubated in 900 μl of APA medium for 3 min. The reaction was then stopped with 2 ml distilled water at a temperature of 0 °C. At this temperature 100 μl of 1 N HCl was added after 1.5 min. After another minute had passed, measurements were performed in semimicrocuvettes (light path 1 cm). Comparative measurements without FBB in the incubation medium were performed during the reaction. Purest-grade FBB (1 mg/ml) was added only after a 3-min reaction time with the 2 ml of distilled water at 0 °C. Because the azo dye that forms is very unstable, it is essential that specified reaction and measuring times be adhered to. Blanks were run parallel to these investigations, i.e., measurements were made without substrate in the incubation medium (measurement of spontaneous azo formation).

APA Measurements with Various Substrates and Buffers. The measurements were made with the APA standard incubation medium. Four different series of measurements were performed: measurements with the normal standard incubation medium (α-L-Glu-MNA as substrate and tris-maleate as buffer); measurements with the standard incubation medium containing α-L-Glu-2NA instead of α-L-Glu-MNA; measurements with the standard incubation medium containing 0.1 M cacodylate buffer, pH 7; and measurements with the standard incubation medium containing 0.1 M cacodylate buffer, pH 7 and α-L-Glu-2NA as substrate. The APA determinations were performed kinetically over 5-min reaction periods in semimicrocuvettes (light path 1 cm).

2.4.3 Fluorometric Measurements of Microdissected Glomeruli (Rat)

Laparotomy and thoracotomy were performed under ether anesthesia, the right atrium was opened, and the animals were perfused for 5 min through the left ventricle with a rinsing solution (cf. Sect. 2.2.2 regarding ultracytochemistry: rinsing solution 1 without procaine and without Thrombophob). The purpose of this rinse was to purge the kidney of plasmatic aminopeptidases. Then the kidneys were quickly removed and frozen as described in Sect. 2.2. These tissue blocks were cut into 35-μm cryostat sections, which were then dried for 24 h in a Leybold-Heraeus apparatus at -50 °C and 10^{-1} to 10^{-2} mmHg using phosphorus pentoxide. Under a stereo microscope the glomeruli were dissected out of the dried tissue sections and collected in Eppendorf reaction vessels. Finally the glomeruli were dissolved in bidistilled water and homogenized. The substrate-dependent APA activities of the glomerular homogenate were measured fluorometrically with the APA fluorometric medium (see above) to determine K_m by the method of Lineweaver-Burk (1934). The enzyme activities were expressed in the form of fluorometric units after reaction periods of 15 or 30 min. In addition to the K_m determination, the type of inhibition produced by ANG I and II (0.025 mM of each in medium) was ascertained using various substrate concentrations (cf. K_m). For control purposes the APM fluorometric medium was also used in place of the APA medium in order to detect any APM activity present in the glomerular preparation.

2.5 Methods of Investigation Kidneys from Experimental Animals (Male Rats and Mice)

Qualitative histochemical studies were carried out with the APA and APM standard incubation media, quantitative histochemical measurements were performed in subcapsular glomeruli with the APA quantifying medium (enzyme-kinetic measurements), and fluorometric measurements were performed in renal homogenate (whole kidneys) to determine APM and APA activities with the corresponding fluorometric media. These investigations were done in all the experimental animals (diet-controlled and adrenalectomized rats and mice) unless otherwise indicated. The kinetically measured fluorometric units were converted to specific enzyme activities (mU/mg protein) by the use of standard curves (cf. Sect. 2.4) and by referring the activities to the protein content of the homogenate (determined by biuret method). The activity changes found in the experimental animals were tested for statistical significance (95% and 99% level) by the U test (statistical program of the Videoplan computer). Incorporated into the U test is a Kolmogoroff-Smirnoff test preset to 90%; binding corrections are routinely made during running of the test.

2.6 Chemicals and Suppliers

Bachem (Bubendorf, Switzerland): L-Ala-2NA; L-Ala-MNA; L-Arg-MNA; α-L-Asp-2NA; α-L-Asp-MNA; ANG I, II, III; α-L-Glu-2NA; α-L-Glu-MNA; L-His-Pro-Phe; L-Leu-MNA; L-Pro-MNA.

Fluka (Neu-Ulm, FRG): Durcupan.

Merck (Darmstadt, FRG): acetone, analytical grade; $CaCl_2$, anhydrous, pure-grade; DMF; maleic acid anhydride for synthesis; NaCl, analytical grade; procaine hydrochloride; tris, analytical grade.

Nordmark (Hamburg, FRG): Thrombophob (5000 U.S.P. units per milliliter sodium heparin).

Riedel-De Haen (Hannover, FRG): D(+)-paraformaldehyde; saccharose.

Roth (Karlsruhe, FRG): cacodylic acid sodium salt trihydrate; glutardialdehyde, 25% puriss. under nitrogen.

Serva (Heidelberg, FRG): EDTA-Na_2 salt, analytical grade; FBB salt, pure grade; FBB · BF_4 salt, purest grade (free of heavy metal); FGGBC salt, pure grade; D-glucose; 2-naphthylamine, purest grade; 1,10-phenanthroline.

Sigma (Munich, FRG): agarose type VII (low gelling temperature); osmium tetroxide.

2.7 Abbreviations

Enzymes: APA = aminopeptidase A; APM = aminopeptidase M; CE = converting enzyme.

Substrates: (cf. IUPAC – IUB Commission on Biochemical Nomenclature, Eur J Biochem 27: 201–207) L-Ala-2NA = L-alanine-2-naphthylamide; L-Ala-MNA = L-alanine-4-methoxy-2-naphthylamide; L-Arg-MNA = L-arginine-4-methoxy-2-naphthylamide; α-L-Asp-2NA = α-L-aspartic acid-2-naphthylamide; α-L-Asp-MNA = α-L-aspartic acid-4-methoxy-2-naphthylamide; α-L-Asp-Arg-MNA = α-L-aspartyl-L-arginine-4-methoxy-2-naphthylamide; α-L-Glu-MNA = α-L-glutamic-acid-4-methoxy-2-naphthylamine; L-Leu-MNA = L-leucine-4-methoxy-2-naphthylamide; L-Pro-MNA = L-proline-4-methoxy-2-naphthylamide.

Coupling agents: FBB = Fast Blue B; FGGBC = Fast Garnet GBC; HNF = hexazotized new fuchsin; HPR = hexazonium-*p*-rosaniline.

Peptides: (cf. Nomenclature of the Renin-Angiotensin System, Klin Wochensch 56 (Suppl I): 187–190, 1978): ANG I = angiotensin I (Asp-Arg-Val-Tyr-Ile-His-Pro-Phe-His-Leu); ANG II = angiotensin II (Asp-Arg-Val-Tyr-Ile-His-Pro-Phe); ANG III = angiotensin III (Arg-Val-Tyr-Ile-His-Pro-Phe); L-His-Pro-Phe = L-histidyl-L-prolyl-L-phenylalanine; L-Tyr-Ile = L-tyrosyl-L-Isoleucine.

Miscellaneous: DMF = *N,N*-dimethylformamide; EDTA = ethylene-diaminetetraacetic acid; FDC sections = freeze-dried, celloidin-mounted cryostat sections; JGA = juxtaglomerular apparatus; K_m = Michaelis constant; K_i = inhibitor constant; RAS = renin-angiotensin system; tris = tris (hydroxymethyl) aminomethane.

3 Results

3.1 Methodological Studies

3.1.1 Qualitative Light-Microscopic Histochemistry

Using the simultaneous azo coupling method, the highest APA activities are demonstrable in un-pretreated cryostat sections with pure- or purest-grade *Fast Blue B* (FBB) or pure-grade *Fast Garnet GBC* (FGGBC) as the coupling agent. However, during and after the reaction there are relatively rapid shifts of the red azo dye with the artifactual formation of granular color products, especially when FGGBC is used (Lojda et al. 1979, Kugler 1981b). Acetone pretreatment of the cryostat sections can largely prevent shifting and granulation of the azo dye from FBB (Fig. 4), though in the case of FGGBC some granulation still occurs (Fig. 5). This is probably due to the adsorption of the azo dye at lipid-water boundaries or its dissolution in lipid droplets. To retain structure when pretreating the sections, it is important that the cryostat sections be fixed in 100% acetone at -20 °C to -30 °C without prior thawing, even though this somewhat decreases the demonstrable enzyme activities (cf. Sect. 3.1.3). When acetone-pretreated sections are used, the following concentrations of the highly water-soluble, stable diazonium salts have proved optimal for the demonstration of APA: pure-grade FBB, 1 mg/ml incubation medium (2.5 mM); purest-grade FBB, 0.5 mg/ml incubation medium (1.1 mM); pure-grade FGGBC, 2–2.5 mg/ml incubation medium (approx. 6.5–8 mM). With lower concentrations of coupling agent, there is an increased diffusion of reaction products (azo dye) into the incubation medium (red staining), especially with FGGBC. At the same time, there is less azo dye formation at the enzyme site in the tissue section with marked shifting of the azo dye in the section. Higher concentrations of coupling agent lead to an increasing inhibition of demonstrable APA acitivites as well as to a dark-red staining of the incubation medium.

The *unstable coupling agents* alternatively used in un-pretreated tissue sections, namely hexazonium-*p*-rosaniline (Fig. 6) and hexazotized new fuchsin, which form a yellow reaction product with 4-methoxy-2-naphthylamine, provide sharp localization but inhibit APA activities more strongly than the stable diazonium salts FBB and FGGBC. Another disadvantage in the use of unstable diazonium salts is that the yellow reaction product contrasts poorly with tissue structures, and reaction sites with low activity are easily overlooked (cf. Lojda and Gossrau 1980, Kugler 1981b).

In working with cryostat sections subjected to various preliminary treatments, we observed that a 5-min *section fixation* at +4 °C in 2% buffered glutaraldehyde or 4% buffered formaldehyde (pH 7.2) caused no significant decrease in APA or APM activities demonstrable by qualitative histochemistry as compared to acetone-pretreated sections. Calcium ions in the fixative do not appear crucial to the maintenance of enzyme activities. Aldehyde pretreatment of the cryostat sections provides a good localization of reaction product compared to un-pretreated cryostat sections. In glutaraldehyde-fixed cryostat sections, the tissue has a nonspecific yellow color following incubation. Compared with un-pretreated cryostat sections, *freeze-dried, celloidin-mounted tissue sections* usually show only low APA activities at reaction times less than 60 min, while the demonstration of APM is less affected by the FDC procedure.

15

An advantage of this procedure is the precise enzyme localization (Fig. 7), although sites with low activity are not well visualized.

Compared with section fixation, the cryostat sections of *block-fixed kidneys* show marked differences between glutaraldehyde and formaldehyde fixation. Fixation with buffered 2% glutaraldehyde for 20 h followed by rinsing in Holt's mixture (1959) for an equal time leads to a very strong decrease in demonstrable APA activities. APM demonstration is considerably less affected by this pretreatment. But on the whole, block fixation in buffered 4% formaldehyde hampers the demonstration of APA and APM a great deal less than in glutaraldehyde, though the demonstrable enzyme activities are still lower than with section fixation in the same fixatives.

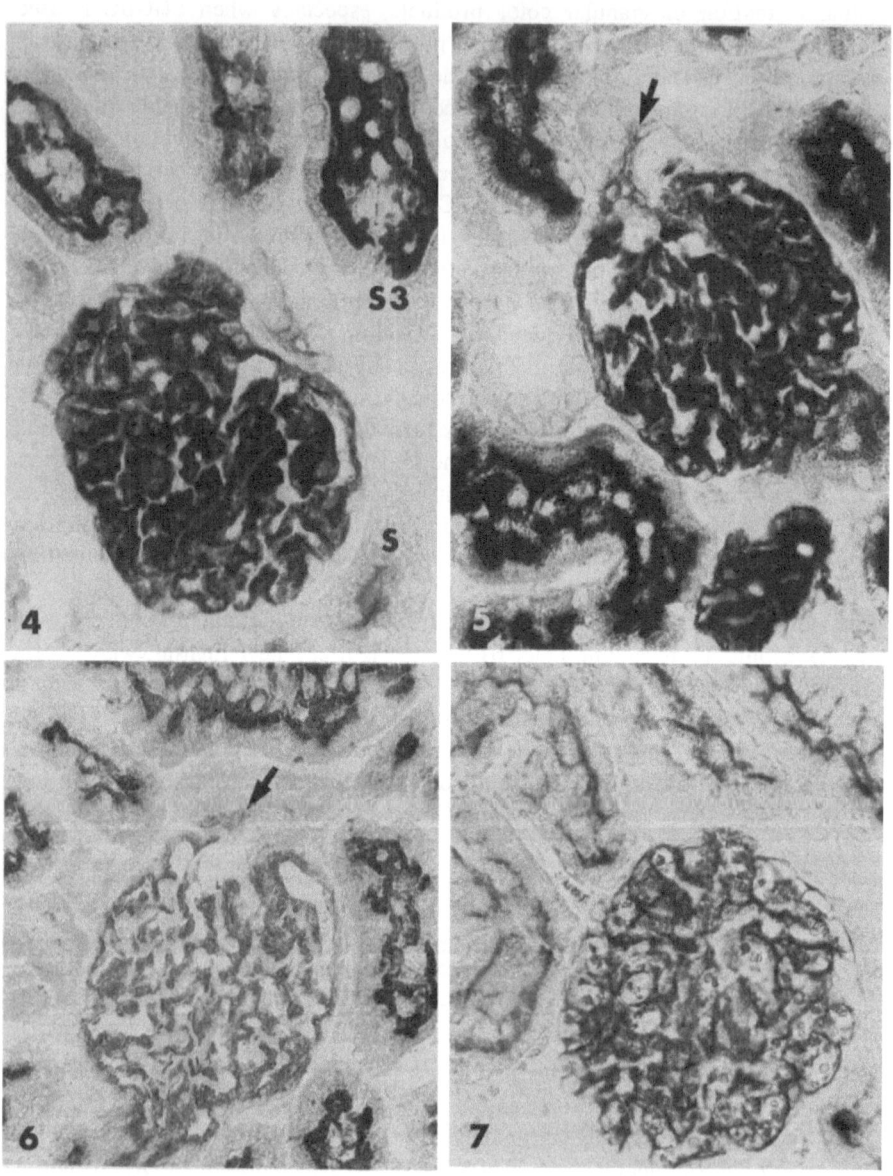

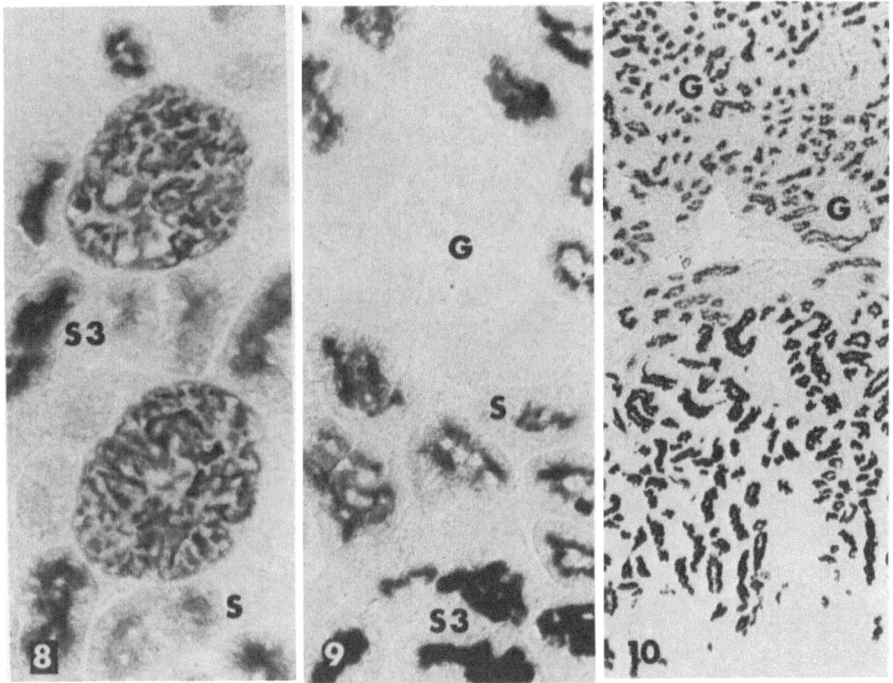

Fig. 8. Male rat. Demonstration of APA. Incubation of 10-μm acetone-pretreated section in APA standard medium for 20 min at +30 °C, containing α-L-Asp-MNA as substrate instead of α-L-Glu-MNA. This substrate is hydrolyzed by APA to a much smaller degree than α-L-Glu-MNA (cf. Fig. 4), resulting in the same reaction pattern with lower demonstrable APA activities. S, S_1 segment; S_3, S_3 segment of proximal tubule. ×140

Fig. 9. Male rat. Demonstration of enzymatic hydrolysis of α-L-Asp-Arg-MNA. Specimen preparation as in Fig. 8 (reaction time 30 min), but with α-L-Asp-Arg-MNA as substrate. This substrate is not attacked in the glomerulus (G), but only at proximal tubule. The reaction pattern thus corresponds to that of APM (cf. Fig. 10). S, S_1/S_2 segment; S_3, S_3 segment of proximal tubule. ×120

Fig. 10. Male rat. Demonstration of APM. Incubation of 10-μm acetone-pretreated section in APM standard medium for 5 min at +30 °C. APM is demonstrable in the brush borders of the proximal tubule but not in the glomerulus (G). ×30

Fig. 4. Male rat. Demonstration of APA. Incubation of 10-μm acetone-pretreated section in APA standard medium (α-L-Glu-MNA as substrate, purest-grade FBB as coupler) for 20 min at +30 °C. The azo dye is demonstrable in the glomerulus and brush border of the proximal tubule (S, S_1/S_2 segment; S_3, S_3 segment). ×300

Fig. 5. Male rat. Demonstration of APA. Specimen preparation as in Fig. 4, but with Fast Garnet GBC (2.5 mg/ml) as coupler. The resulting azo dye is partially granular. Reaction sites as in Fig. 4. An APA-positive Goormaghtigh's cell nest is also present (*arrow*). ×300

Fig. 6. Male rat. Demonstration of APA. Incubation of 10-μm un-pretreated tissue section in APA standard medium for 30 min at +30 °C containing 0.025 ml/ml HPR instead of FBB. APA is demonstrable at the same sites as in Fig. 5, but with lower activities. *Arrow*, Goormaghtigh's cell nest. ×275

Fig. 7. Male rat. Demonstration of APA. Incubation of 10-μm FDC section in APA standard medium for 60 min at +30 °C. Localization is sharp, but APA reaction rates in the glomerulus and brush borders of the proximal tubule are low (long incubation time). ×300

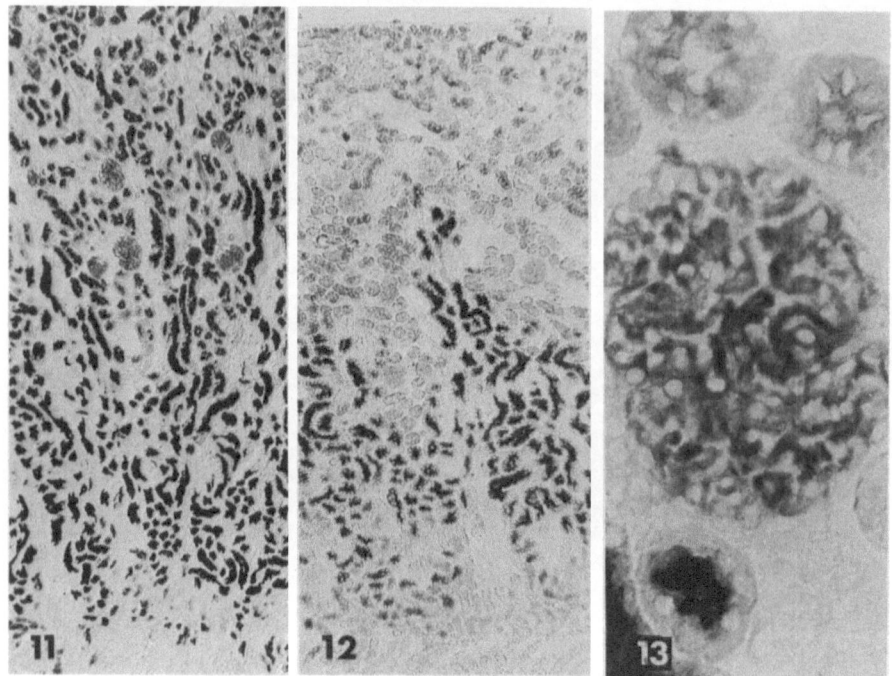

Fig. 11. Male rat. Demonstration of APA. Specimen preparation as in Fig. 8 but with α-L-Glu-MNA as substrate. APA is demonstrable in glomeruli and brush borders of the proximal tubule (especially S_3). ×25

Fig. 12. Male rat. Demonstration of APA. Specimen preparation as in Fig. 4. Instead of $CaCl_2$, the incubation medium contains 5 mM EDTA as a complexing agent for divalent cations. A strong decrease in demonstrable APA activities is observed at all reaction sites (cf. Fig. 11). ×30

Fig. 13. Male rat. Demonstration of APA. Enlarged section from Fig. 12. Low APA activities in glomerulus and S_1 segment of proximal tubule (cf. Fig. 4). ×275

Fig. 14. Male rat. Demonstration of APM. Incubation of 10-μm acetone-pretreated section in APM standard medium for 10 min at +30 °C. APM is found in the brush borders of the proximal tubule, especially in S_3 segment (strong reaction of subcortical zone). ×25

Fig. 15. Enlarged section from Fig. 14. Glomeruli (G) are devoid of APM. ×100

Fig. 16. Male rat. Demonstration of APM. Specimen preparation as in Fig. 14, but with 5 mM $CaCl_2$ in incubation medium. Reaction pattern and APM activities same as without $CaCl_2$ (cf. Fig. 14). ×25

Fig. 17. Enlarged section from Fig. 16. G, glomerulus. ×100

Fig. 18. Male rat. Demonstration of APM. Specimen preparation as in Fig. 14, but with 5 mM EDTA in incubation medium. Reaction pattern and APM activities same as without EDTA (cf. Fig. 14). ×25

Fig. 19. Enlarged section from Fig. 18. G, glomeruli. ×100

18

When *various substrates* are used in the APA incubation medium, it is found that simultaneous azo coupling with α-L-Asp-MNA yields the same activity pattern as α-L-Glu-MNA (Fig. 8). The enzymic reaction rate is much lower, however. α-L-Glu-2NA, the substrate most commonly used by other investigators (Glenner and Folk 1961, Glenner et al. 1962, Wachsmuth and Donner 1976), leads to an increase in demonstrable APA acitivities, but also to a very imprecise localization of reaction products in the tissue section, to a relatively strong azo dye formation in the incubation medium, and to a diffuse precipitation of azo dyes on the tissue section. A more precise enzyme localization with this substrate cannot be obtained with the

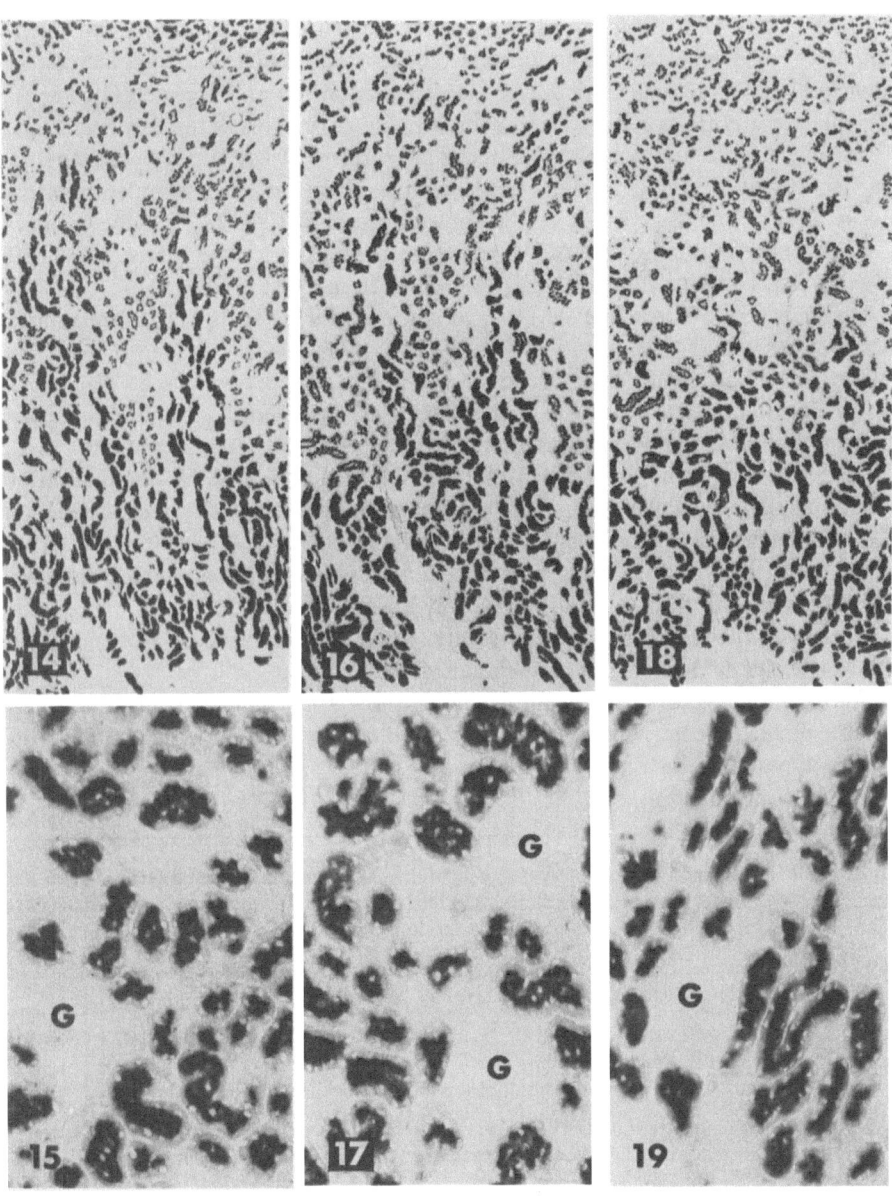

histochemical procedure described. In the demonstration of both APA and APM, the use of 4-methoxy-2-naphthylamide derivatives leads to a considerable improvement of localization (cf. Lojda et al. 1979, Lojda and Gossrau 1980). With the substrates α-L-Asp-Arg-MNA (Fig. 9), L-Ala-MNA (Fig. 10), L-Arg-MNA, and L-Leu-MNA, similar results are obtained on incubation with the APA and APM medium (see below), i.e., results correspond to the APM reaction pattern (Fig. 10). It is noteworthy that the presence of 2.5 mM calcium chloride significantly increased the enzymic attack of α-L-Asp-Arg-MNA. The L-Pro-MNA used as a substrate in the APM incubation medium is not demonstrably attacked in the tissue section.

The APA and APM incubations run without substrate for control purposes give negative results.

In APA und APM demonstrations using incubation media of varying *ionic composition*, we made the following findings: Calcium ions lead to a strong increase in demonstrable APA activities; as calcium ion concentrations are increased (1.5, 5, 10 mM), there is a corresponding rise in demonstrable enzyme activities (cf. Sect. 2.3). EDTA, which forms ion complexes, inhibits the enzyme (Figs. 12, 13) compared to controls without EDTA (Fig. 11). APM, on the other hand, shows no qualitative histochemical differences relative to the control reactions (Figs. 14, 15) under different incubation conditions; i.e., APM is neither activated by calcium ions (1.5, 5, 10 mM) (Figs. 16, 17) nor inhibited by EDTA (1.5, 5, 10 mM) (Figs. 18, 19).

Common to both enzymes is their very strong inhibition by 1,10-phenanthrolene.

Further important findings are obtained when substrate-equimolar concentrations of *angiotensins* and *smaller peptides* (peptide fragments of angiotensins) are added to the APA and APM incubation media. Specifically, it is found that ANG I, II, and III strongly inhibit both APA (Figs. 20, 21) and APM. A comparison of ANG II and III reveals that ANG II is the stronger inhibitor of APA, and ANG III of APM. The peptides L-Tyr-Ile (Fig. 22) and L-His-Pro-Phe (Fig. 23) also inhibit APM, but much less than the angiotensins (cf. Sect. 2.4.1). L-Tyr-Ile has no inhibitory effect on APA demonstration, and L-His-Pro-Phe only a slight effect.

When *cacodylate buffer* is used for the APA reaction instead of tris-maleate buffer, somewhat higher enzyme activities are demonstrated.

The methodological histochemical studies show that shorter incubation times generally provide clearer, more informative results due to the fact that longer incubation times obscure activity differences (cf. Sect. 3.1.3, time-dependent APA measurements in glomeruli).

Fig. 20. Male rat. Demonstration of APA. Incubation of 10-μm acetone-pretreated section for 20 min at +30 °C in APA standard medium also containing substrate-equimolar amounts of ANG II. ANG II produces strong inhibition at all APA reaction sites. ×30

Fig. 21. Enlarged section from Fig. 20. Glomeruli (*arrows*) show low APA activities compared to demonstrations in ANG II-free reaction media (cf. Fig. 4). ×160

Fig. 22. Male rat. Demonstration of APM. Incubation of 10-μm acetone-pretreated section for 5 min at +30 °C in APM standard medium also containing substrate-equimolar amounts of L-Tyr-Ile. This peptide causes a relatively strong inhibition of APM localized in the brush borders. ×30

Fig. 23. Male rat. Demonstration of APM. Specimen preparation as in Fig. 22, but with the peptide L-His-Pro-Phe instead of L-Tyr-Ile in the incubation medium. The APM demonstration corresponds to Fig. 22 in localization and activity. ×30

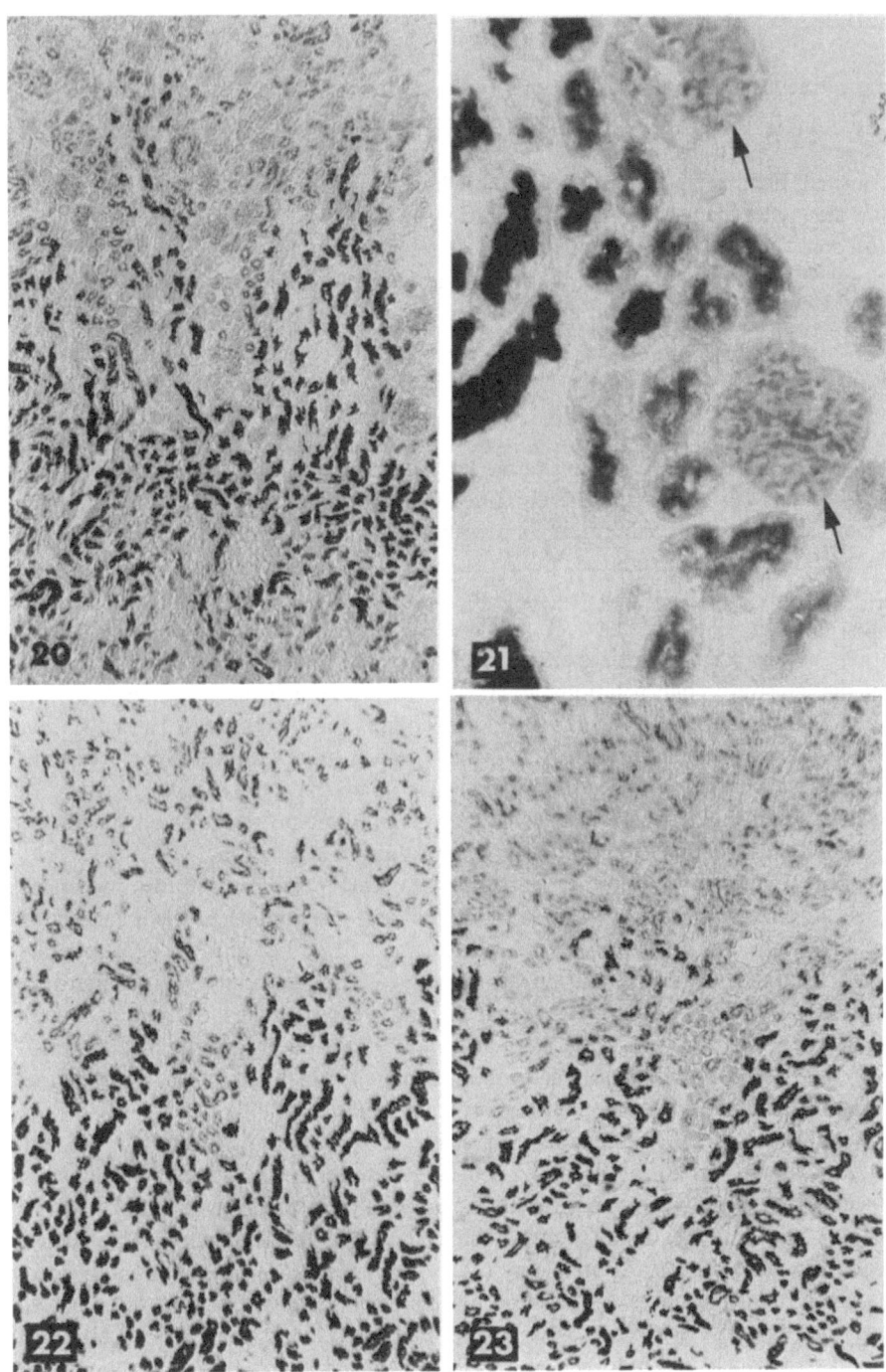

3.1.2 Ultracytochemistry

In the ultracytochemical investigations of APA, it was found that when tissue pieces from perfusion-fixed kidneys are incubated, reactions are confined mainly to the superficial portions of the tissue, apparently because the incubation components have a small depth of penetration. On the whole, perfusion fixation with low-concentration aldehydes leads to a marked inhibition of APA, which is in contrast with light-microscopic findings. APA demonstration with hexazonium-*p*-rosaniline as the coupling agent yields the best results at pH 5. The control incubations without substrate lead to no formation of reaction product.

3.1.3 Quantitative Histochemistry

Purest-grade FBB is used for quantitative histochemical studies, since, unlike FGGBC, this diazonium salt yields a relatively amorphous reaction product (cf. Sect. 3.1.1), and there is less enzyme inhibition than with HPR or HNF (cf. Sect. 3.1.1; Lojda and Gossrau 1980). One characteristic of the substrates α-L-GluMNA and α-L-Asp-MNA is their very low water solubility. Complete dissolution of the substrates in DMF is possible only by alkalization (cf. Lojda and Gossrau 1980). We use 0.4 M tris stock solution for this purpose. This makes it possible to keep the DMF concentrations low. When the substrate concentrations are varied, we use equalizing solutions to ensure that the DMF concentration in the reaction medium is always equal to 1.8% v/v. Thus our DMF concentrations are slightly above 1%, which reportedly has no effect on the APA reaction rate (Hess 1965). It should also be noted that our reaction solutions for quantitative and qualitative enzyme histochemistry show no flocculation (no filtration), but retain a clear yellow color even with prolonged incubation.

Before kinetic APA activity measurements were performed, it was necessary to determine the *absorption spectrum* of the red azo dye formed in the renal section from 4-methoxy-2-naphthylamine and purest-grade FBB. For this purpose reaction sites with strong and weak activity in the kidney were measured microdensitometrically between 470 and 600 nm. Both the strong and weak sites show a common absorption maximum at 500 nm (Fig. 24). In determining the absorption maximum of tissue-bound azo dye, aftertreatment of the section with aldehydes must be avoided, as they produce a change in color quality and thus in absorbance.

Because APA is an ion-dependent enzyme (McDonald and Schwabe 1979, Kenny 1979; Lojda and Gossrau 1980; Kugler 1981b lit.), we use as our quantifying medium an incubation solution containing only 1.5 mM calcium chloride (unless otherwise indicated) and no other ions (sodium, chloride, etc.) — at least not in appreciable concentrations. This is made possible by the use of purest-grade FBB as coupling agent and tris-maleate solution as buffer. For the enzyme-kinetic measurements discussed below, only subcapsular glomeruli were used. Because the densitometric measurement of APA activities is followed by morphometric analysis of the reaction area, it is essential that structural detail be preserved and reaction products well localized (cf. Sect. 3.1.1). Therefore, the tissue sections are fixed in acetone for 2 min at −25 °C prior to incubation. A comparison with kinetic measurements in tissue sections not treated with acetone, however, showed that APA activities are decreased by about 13% after acetone pretreatment.

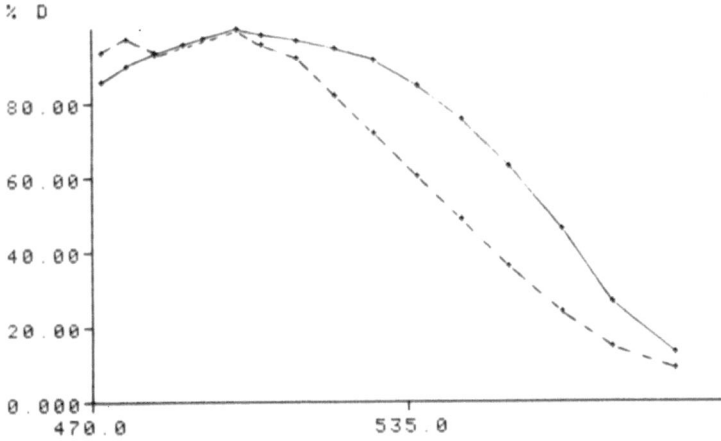

Fig. 24. Determination of the absorption spectrum of the azo dye from 4-methoxy-2-naphthylamine and purest-grade FBB in the renal section. 10-μm cryostat sections pretreated in acetone were incubated in the APA standard medium for 10 min at +30 °C, then washed for 10 min in distilled water at +4 °C and mounted in glycerin gelatin. Reaction sites in the kidney with weak (– – –) and strong activity (——) were measured with the M 85 microdensitometer between 470 and 600 nm. M 85 settings: bandwidth 15, spot size 2, scanning frame size 1, gating mask A1, objective 40×. The absorption peak occurs at 500 nm. *NM*, wavelength in nm; *%D*, percentage density distribution along the absorption spectrum; the maximum density at 500 nm was set equal to 100%

Our enzyme-kinetic measurements (determinations of initial enzyme activities) *as a function of substrate concentration* indicate that with α-L-Glu-MNA as substrate and 1.5 mM calcium chloride (pH 7) at a reaction temperature of +30 °C and section thickness of 5 μm, the K_m for APA is equal to 0.23 mM in rat glomeruli (corrected for spontaneous azo dye formation; the uncorrected value is 0.16 mM) and 0.14 mM in mouse glomeruli (uncorrected for spontaneous azo formation) (Figs. 28–30, Table 1). The V_{max} in mouse glomeruli is 1.6 times higher than in rat glomeruli.

APA measurements *as a function of reaction time* (+30 °C reaction temperature and 5-μm section thickness) show a linear relationship between reaction time and formation of reaction product only during the first 1–2 min of the reaction (Fig. 25). This finding demonstrates the importance of the measurement of initial enzyme activities appling quantitative histochemical methods.

The amount of reaction product formed *as a function of section thickness* represents a key quantitative histochemical parameter (Wachsmuth 1980; Kugler 1981a). Our kinetic measurements in rat glomeruli show a linear relationship between the formation of reaction product and section thickness between 3 and 12 μm, with the regression line passing through the origin (Fig. 26).

The determination of APA activities under *various ionic conditions* (Fig. 27) shows a clear calcium-ion dependence of APA, its activity increasing with the concentration of CaCl$_2$ in the incubation medium. Thus, the addition of 10 mM CaCl$_2$ to the medium leads to a 2.5-fold increase in APA activity. The use of 1.5 mM EDTA produces a slight decrease in APA activities compared to the calcium chloride-free incubation medium. Higher EDTA concentrations may not be used in the quantifying medium, as they would interfere with accurate morphometric analysis.

Table 1. Quantitative histochemical (glomeruli) and biochemical (renal homogenate and glomeruli) data on APA and APM (male rats)[a]

	Procedure	Biochemistry (fluorometry)	Biochemistry (fluorometry)	Quantitative histochemistry (microdensitometry)
	Demonstration site	Renal homogenate	Glomeruli	Glomeruli
APA	Km	0.130 mM	0.230 mM	0.230 mM
	ANG I	Noncompetitive K_i = 0.015 mM	Competitive	Competitive
	ANG II	Competitive K_i = 0.015 mM	Competitive	Competitive K_i = 0.140 mM
	ANG III	Noncompetitive K_i = 0.040 mM	–	Competitive
APM	Km	0.240 mM	–	–
	ANG I	Noncompetitive K_i = 0.045 mM	–	–
	ANG II	Noncompetitive K_i = 0.075 mM	–	–
	ANG III	Competitive K_i = 0.003 mM	–	–
APA/APM		1/1.6	6/1	69/1

[a] K_m, the types of inhibition caused by ANG I, II, and III, and the resulting K_i were determined. The K_m and inhibition type were determined from the Lineweaver-Burk plot (1934), and K_i from the secondary Lineweaver-Burk plot (cf. Bisswanger 1979). In each case K_i was determined for at least two different angiotensin concentrations. The fluorometric measurements were performed with L-Ala-2NA (APM) and α-L-Glu-2NA (APA). Simultaneous azo coupling with α-L-Glu-MNA or L-Ala-MNA and purest-grade FBB were used for the histochemical measurements. The APA reaction media always contained 1.5 mM CaCl$_2$. The reaction pH was 7.0 in both enzyme demonstrations; the reaction temperature for fluorometric measurements was +25 °C and +30 °C for microdensiometric measurements.

The influence of NaCl on glomerular APA activities was also investigated, because it was suspected that NaCl may be responsible for the differences in APA activity associated with the use of different buffers with varying sodium and chloride contents (cf. Lojda and Gossrau 1980). Our measurements indicate that APA activities increase with the concentration of NaCl. Thus, when 100 or 130 mM NaCl is added to the incubation medium, 16%–17% higher enzyme activities are measured than without NaCl. Higher NaCl concentrations lead to a decrease in measurable APA activities,

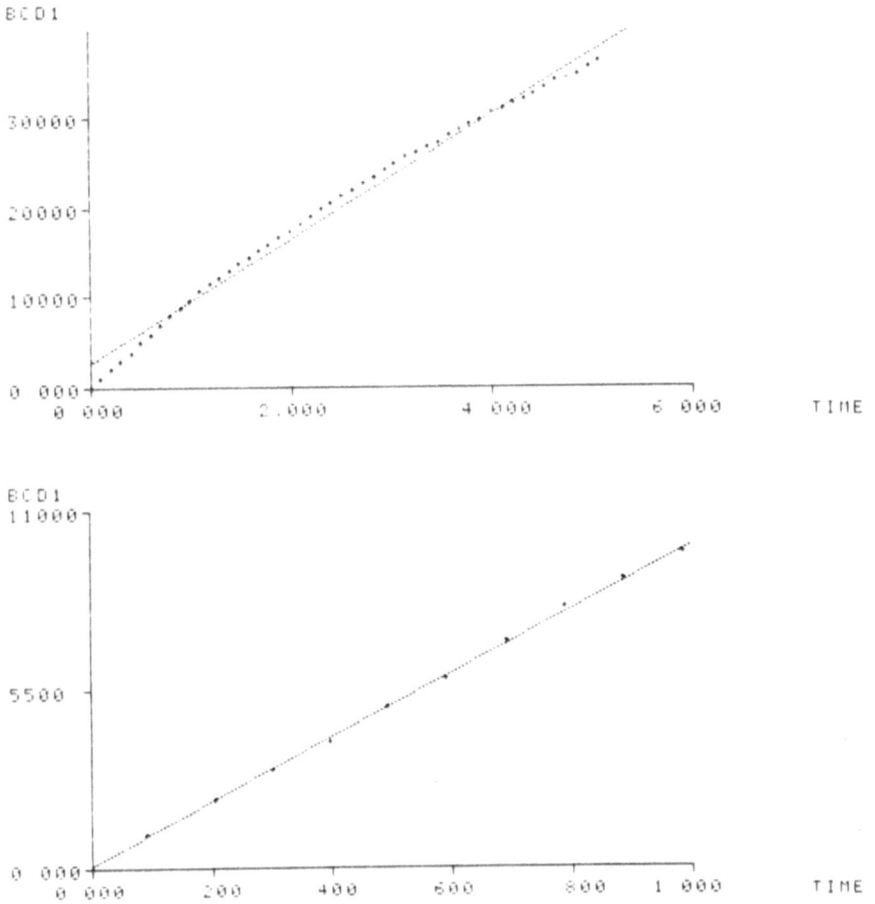

Fig. 25. Quantitative histochemical APA measurements in subcapsular glomeruli of the rat as a function of reaction time (reaction temperature +30 °C, 5-μm sections). Kinetic measurements were made at 6-s intervals over a 5-min period using the APA quantifying medium (*upper graph*). The classification program (linear regression analysis) of the Videoplan computer indicates a linear relation between reaction time and formation of reaction product ($r = 0.999$) only during the initial 1–2 min (*lower graph*). *TIME*, reaction time in min; *BCD 1* relative density values measured at 500 nm.

however, with 180 mM NaCl in the medium producing a 5% inhibition. These findings may explain the approximately 14% higher APA activities observed with the use of 0.1 M sodium cacodylate buffer than with tris-maleate buffer.

Our kinetic measurements in glomeruli with *different substrates* indicate a 40% decrease in APA activities when α-L-Glu-MNA is replaced by α-L-Asp-MNA (Fig. 27), which, according to qualitative histochemical findings, is split at the same sites as α-L-Glu-MNA (cf. Sect. 3.1.1). When α-L-Glu-MNA is replaced with L-Ala-MNA, which is a substrate of APM − an enzyme not demonstrable in the glomerulus according to qualitative histochemical findings (cf. Sect. 3.2) − only very low reaction rates are measurable. The enzyme activities demonstrable with L-Ala-MNA are about 1.4% of those with α-L-Glu-MNA, thus implying a 69:1 ratio of APA to APM (Table 1). It

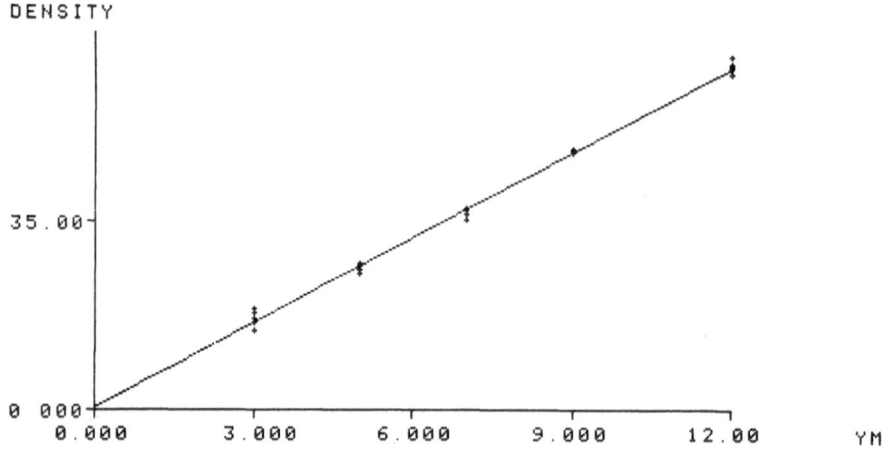

Fig. 26. Formation of reaction product by APA activities (quantitative histochemical measurements) in subcapsular glomeruli as a function of section thickness (3–12 μm); reaction temperature +30 °C. Initial enzyme activities were measured with the APA quantifying medium; at least five measurements were performed per section thickness. The classification program of the Videoplan computer showed a linear relation (r = 0.999) between relative density values (at 500 nm) per min per μm² reaction area (= *Density*) and section thickness (= *YM*), the regression line passing through the origin of the graph

should be noted that the kidneys used for these measurements were not perfused with rinsing solution to eliminate blood constituents that might contain APA and APM.

Using fixed *angiotensin concentrations* with various substrate concentrations (cf. procedure of Hess 1965, Bisswanger 1979), we determined the types of inhibition produced by ANG I, II, and III in the glomerular APA of the rat and mouse (Figs. 28–30, Table 1). The Lineweaver-Burk plot indicates a competitive inhibition for all three angiotensins, with the following sequence of inhibitory intensity: ANG I > ANG III > ANG II. The K_i with ANG II determined for APA in rat glomeruli is 0.14 mM (corrected for spontaneous azo dye formation).

Densitometric *control measurements* in glomeruli using an incubation medium devoid of substrate in order to determine spontaneous azo dye formation indicate that this azo formation accounts for about 11% of the kinetically measured APA activities (using a quantifying medium with substrate) at a reaction temperature of +30 °C. This spontaneous azo dye formation takes place in the tissue section and in the overlying layer of incubation medium (200 μm thickness defined by spacer wire). If the spontaneous azo dye formation is measured only in the layer of reaction medium (without a tissue section), it equals about 33% of the total spontaneous azo dye formation takes place within the section.

Our quantitative histochemical findings on the *APA activity pattern along the nephron* (Fig. 31) are in agreement with qualitative histochemical results (cf. Sect. 3.2). The highest APA activities are found in the S_3 segment of the proximal tubule, and the lowest in the S_1 segment. If we set the APA activities in the brush border of segment S_3 equal to 100%, then the activities are approximately 63% in S_2, 14% in S_1, and 31% in the glomerulus. This applies only to male rats, however. When substrate-equimolar ANG II is added to the APA incubation medium, there is a strong

26

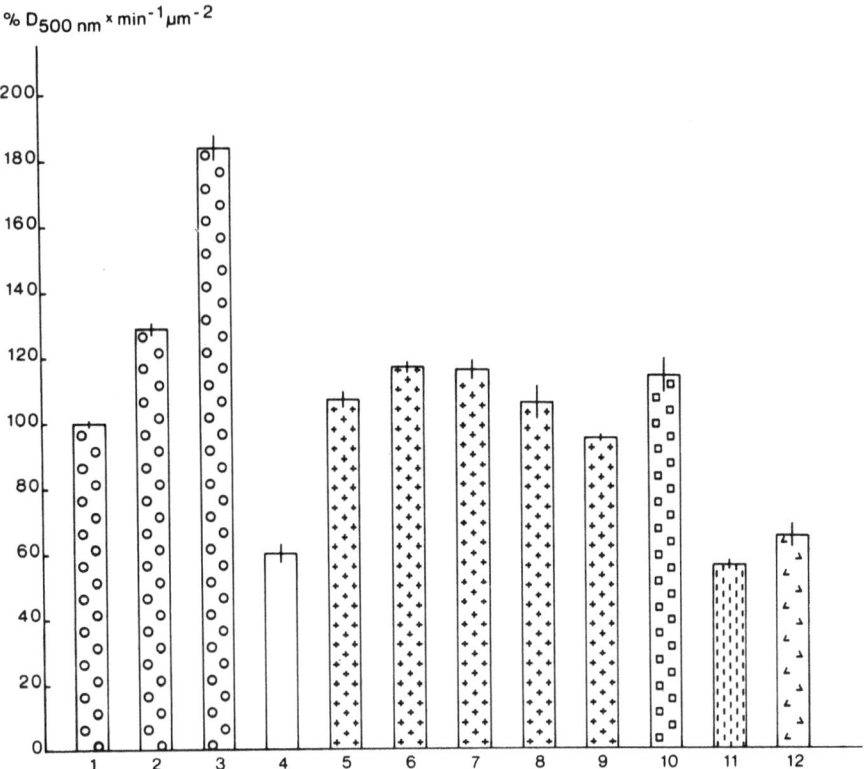

% D$_{500\,nm}$ × min^{-1} μm^{-2}

Fig. 27. Determination of APA activities in subcapsular glomeruli of rats (quantitative enzyme histochemistry) under various ionic conditions (*columns 1–11*) and with the substrate α-L-Asp-MNA (*column 12*). Kinetic measurements of initial enzyme activities (*D*, relative density values) were performed with the APA quantifying medium (reaction temperature +30 °C, 5-μm sections). At least five measurements were performed for each of the different incubation conditions. *Column 1* = 1.5 mM CaCl$_2$; *2* = 5 mM CaCl$_2$; *3* = 10 mM CaCl$_2$; *4* = without CaCl$_2$; *5* = 50 mM NaCl; *6* = 100 mM NaCl; *7* = 130 mM NaCl; *8* = 150 mM NaCl; *9* = 180 mM NaCl; *10* = 0.1 M sodium cacodylate buffer instead of tris-maleate buffer, *11* = 1.5 mM EDTA; *12* = mM α-L-Asp-MNA instead of α-L-Glu-MNA. The APA activities of columns 2–12 were set in relation to column 1 (= 100%). Description of results in text

inhibition of APA activities in the glomerulus and S$_3$ segment (no measurements could be made in S$_1$ and S$_2$ due to differentiation problems under these conditions), yet the inhibition rate is equal at both sites (60%).

3.1.4 Biochemistry of APA and APM

Our biochemical studies of APA and APM consisted mainly of kinetic fluorometric measurements using 2-naphthylamide derivatives. Our first step therefore was to prepare serial dilutions of buffered 2-naphthylamine solutions to ensure that a linear relationship exists between the 2-naphthylamine concentrations used and the measured fluorescence (Fig. 32). The 2-naphthylamines were investigated in a concentration range corresponding to the quantities of 2-naphthylamine liberated by enzyme

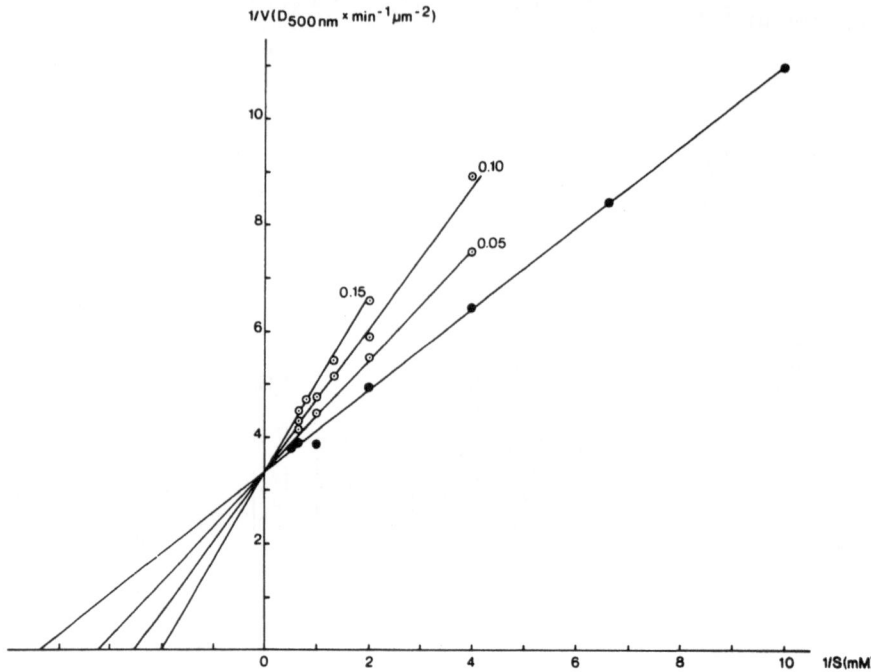

Fig. 28. APA activities plotted against various substrate concentrations ($\bullet$, $r = 0.980$) and ANG II concentrations ($\odot = 0.05$ mM, $r = 0.996$; 0.1 mM, $r = 0.993$; 0.15 mM, $r = 0.996$) by the method of Lineweaver and Burk (1934). Initial APA activities measured in subcapsular rat glomeruli by quantitative histochemical methods. 5-μm acetone-pretreated renal sections were incubated, with the quantifying medium (simultaneous azo coupling; $0.1-1.5$ mM α-L-Glu-MNA as substrate and 0.5 mg/ml purest-grade FBB as coupling agent) at +30 °C. The values plotted are corrected for spontaneous azo dye formation. Each data point represents the mean of at least five series of measurements. The K_m is 0.23 mM. ANG II leads to a competitive inhibition of APA and a K_i 0.14 mM

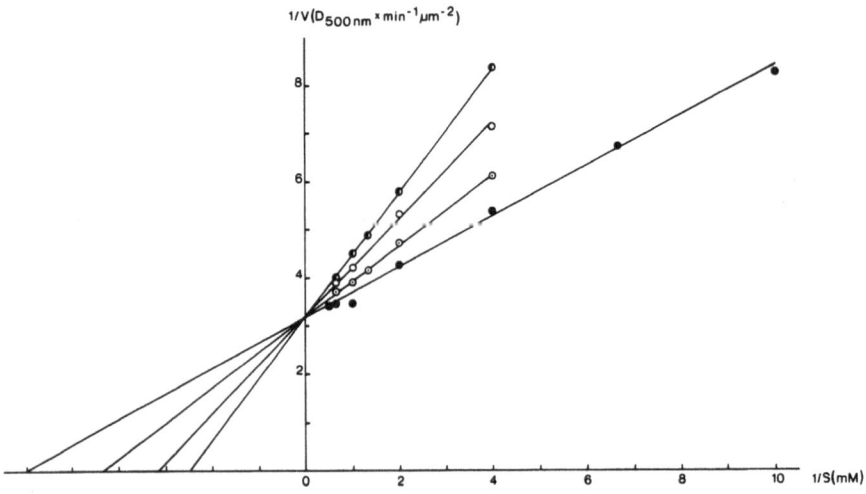

Fig. 29. APA activities plotted against various substrate concentrations ($\bullet$) and angiotensins ($\copyright$ = ANG I, $\odot$ = ANG II, $\bigcirc$ = ANG III, each 0.05 mM) by the method of Lineweaver and Burk (1934). Quantitative histochemistry of subcapsular glomeruli of the rat. Procedure as in Fig. 28. All angiotensins competitively inhibit APA

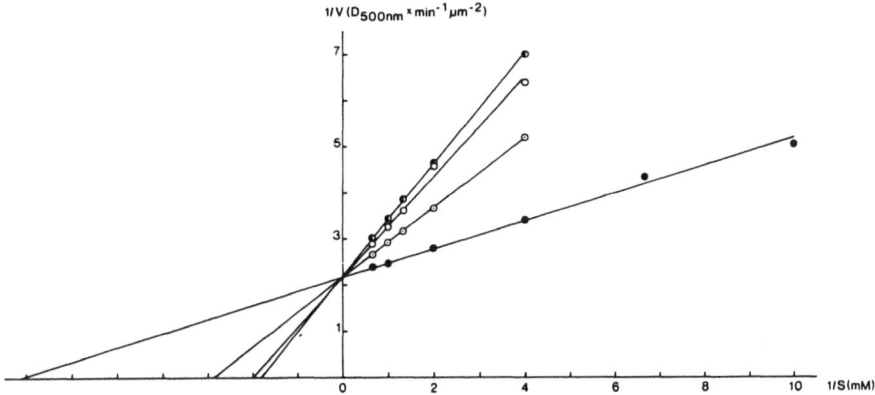

Fig. 30. APA activities plotted against various substrate concentrations ($\bullet$) and angiotensins ($\mathbb{O}$ = ANG I, $\odot$ = ANG II, $\bigcirc$ = ANG III, each 0.05 mM) by the method of Lineweaver and Burk (1934). Quantitative histochemistry of subcapsular glomeruli of the mouse. Procedure as in Fig. 28. All angiotensins competitively inhibit APA

activities. The standard curves for 2-naphthylamines in tris-HCl buffer and cacodylate buffer are nearly identical.

For the *reaction solutions*, all substrates were dissolved in DMF and 0.4 M tris stock for comparison purposes (L-Ala-2-NA is also soluble in DMF alone); the final concentration of DMF in the reaction medium was always 1.6% v/v. The only buffer used for fluorometric studies (except for the substrate-dependent APA measurements) was cacodylate buffer (cf. Lojda and Gossrau 1980). Tris-maleate is unsuitable because of its interaction with 2-naphthylamine, and tris-HCl due to its unstable buffering action in the pH range employed.

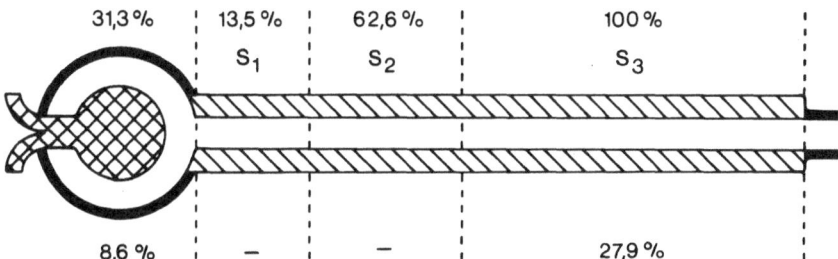

Fig. 31. APA distribution pattern along the rat nephron (quantitative histochemical endpoint measurements). 7-μm cryostat sections were fixed in 100% acetone for 2 min at $-25\,°$C and then incubated in APA standard medium for 1.5 min at $+30\,°$C. After the reaction the sections were washed in acidic distilled water (50 ml distilled water and 1.35 ml of 1 N HCl; pH 2) for 5 min at $0\,°$C and mounted in Karion F. The APA activities along the nephron were measured against APA-free regions of the renal medulla (0 values) (M 85 settings as in Fig. 24). The series of numbers above the schematically drawn nephron indicate the relative APA distribution (highest activities in S_3 = 100%). In addition, the APA activities in the glomerulus and S_3 segment were measured after incubation with substrate-equimolar ANG II added to the reaction medium (numbers below the nephron). ANG II causes a 60% inhibition of APA in the nephron regions investigated. S_1, S_2, S_3 = segments of the proximal tubule

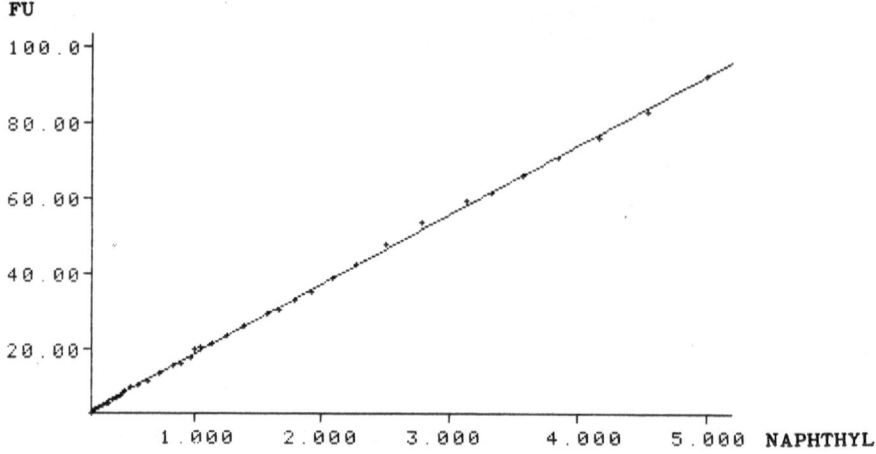

Fig. 32. Standard curve for determining the relation between various 2-naphthylamine concentrations and fluorescence. For this purpose aliquots of 0.005–0.125 μmol 2-NA taken from a 2-NA stock (14.3 mg dissolved in 4 ml DMF) were dissolved in 0.1 M sodium cacodylate buffer, pH 7. The fluorescence of the different 2-NA concentrations was determined with a ratio fluorometer (Farrand Foci, primary filter 363 nm, secondary filter 450 nm). The classification program of the Videoplan computer was used to determine the relation between 2-NA concentration (*Naphthyl*; 1.000 = 0.025 μmol 2-NA) and fluorometric units (*FU*) (r = 1.00)

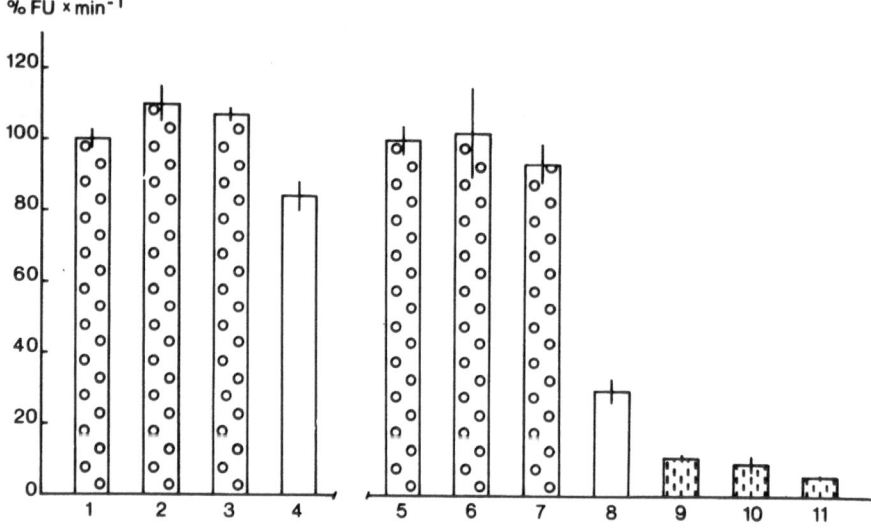

Fig. 33. Determination of APA activities in the renal homogenate of the rat (biochemical fluorometric measurements) under various ionic conditions. The kinetic measurements of initial enzyme activities (*FU*, fluorometric units) were performed with the APA fluorometric medium in the ratio fluorometer (specifications in Fig. 32; reaction temperature +25 °C). At least five measurements were performed for each of the different incubation conditions. *Columns 1–4*: 1.5 mM α-L-Glu-2NA as substrate; *columns 5–11*: α-L-Asp-2NA as substrate. *Columns 1 + 5* = 1.5 mM CaCl$_2$; *2 + 6* = 5 mM CaCl$_2$; *3 + 7* = 10 mM CaCl$_2$; *4 + 8* = without CaCl$_2$; *9* = 1.5 mM EDTA; *10* = 5 mM EDTA; *11* = 10 mM EDTA. The APA activities of columns 2–4 were set in relation to column 1 (= 100%), and those of columns 6–11 to column 5 (= 100%). Description of results in text

30

Our measurements *as a function of substrate concentration* were performed with both cacodylate and tris-HCl buffer in the case of APA, and only with cacodylate in the case of APM. These measurements in renal homogenate indicate a K_m of 0.13 mM for APA with α-L-Glu-2NA as substrate and 1.5 mM CaCl$_2$ with both cacodylate and tris-HCl buffer, and a K_m of 0.24 mM for APM with L-Ala-2NA as substrate and cacodylate buffer (reaction temperature +25 °C, pH 7) (Table 1, Figs. 35, 36). The V_{max} of APA is 70% lower than that of APM. Comparing the activities of the two enzymes, we find a 1:1.6 ratio of APA to APM (Table 1).

The fluorometric APA measurements in renal homogenate using reaction media of varying *ionic composition* indicate that calcium ions activate APA up to a concentration of 5 mM (Fig. 33). If we compare the two substrates of APA (α-L-Asp-2NA displaying a 20% lower reaction rate than α-L-Glu-2NA), we find that the calcium effect is substantially greater when α-L-Asp-2NA is used. Thus the use of 1.5 mM calcium chloride with α-L-Asp-2NA as substrate more than triples the measurable APA activities. With both substrates, a decline in demonstrable APA activities is seen when the calcium chloride concentration is increases to 10 mM. The effects of EDTA are also very pronounced in measurements with α-L-Asp-2NA as substrate. Thus, 1.5 mM EDTA decreases enzyme activity by about 65%, and by about 90% compared to measurements with 1.5 mM calcium chloride. Increasing the EDTA concentration to 10 mM leads to a further slight decrease in APA activities (Fig. 33).

APM behaves differently under these ionic conditions (Fig. 34). With up to 5 mM calcium chloride in the incubation medium, no clear changes in APM acti-

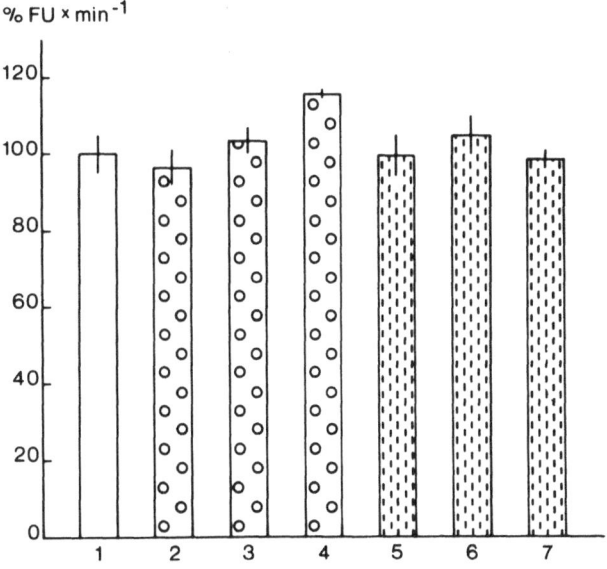

Fig. 34. Determination of APM activities in the renal homogenate of the rat (biochemical fluorometric measurements) under various ionic conditions. The kinetic measurements of initial enzyme activities (*FU*, fluorometric units) were performed with the APA fluorometric medium in the ratio fluorometer (specifications in Fig. 32; reaction temperature +25 °C, *L*-Ala-2NA as substrate). At least five measurements were performed for each incubation condition. *Column 1* = without CaCl$_2$; *2* = 1.5 mM CaCl$_2$; *3* = 5 mM CaCl$_2$; *4* = 10 mM CaCl$_2$; *5* = 1.5 mM EDTA; *6* = 5 mM EDTA; *7* = 10 mM EDTA. The APM activities of columns 2–7 were set in relation to column 1 (= 100%). Description of results in text

vities can be demonstrated. With 10 m*M* calcium chloride, however, there is an apparent increase in APM activities (about 15%). EDTA causes no marked changes in concentrations of 1.5, 5, and 10 m*M*.

The *inhibition experiments with angiotensins* (I, II, and III) in renal homogenate (Figs. 35, 36, Table 1) indicate that ANG II competitively inhibits APA according to the Lineweaver-Burk plot with a K_i of 0.015 m*M*, while ANG III competitively inhibits APM with a K_i of 0.003 m*M*. Inhibition of APA by ANG I and III, and of APM by ANG I and II is noncompetitive in nature. A further important finding is that APM is inhibited 25% by 1.5 m*M* *L-Tyr-Ile* and *L-His-Pro-Phe* (peptide fragments of angiotensins), and APA to an equal degree by L-His-Pro-Phe (Fig. 37). L-Tyr-Ile has no pronounced effect on APA.

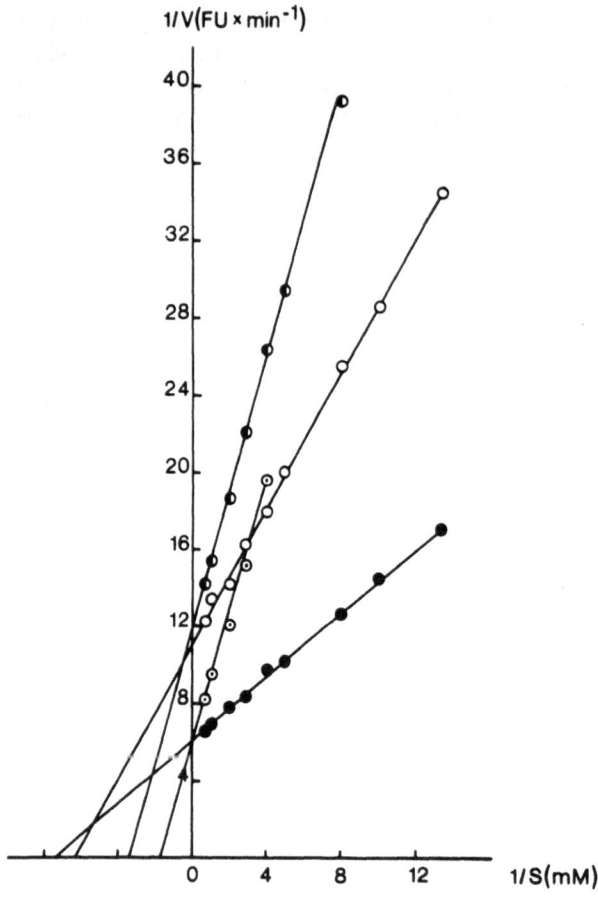

Fig. 35. APA activities plotted against various substrate concentrations (●, *r* = 0.990) and angiotensins (◐ = 0.05 m*M* ANG I, *r* = 0.937; ☉ = 0.05 m*M* ANG II, *r* = 0.982; ○ = 0.05 m*M* ANG III, *r* = 0.989) by the method of Lineweaver and Burk (1934). The APA activities were kinetically determined with the APA fluorometric medium in the ratio fluorometer (specifications in Fig. 32; *FU*, fluorometric units) with 0.075–1.5 m*M* α-L-Glu-2NA in the renal homogenate (rat) (reaction temperature +25 °C). Each data point represents the mean of at least five series of measurements. The K_m is 0.13 m*M*. ANG II produces a competitive inhibition, and ANG I and III a noncompetitive inhibition

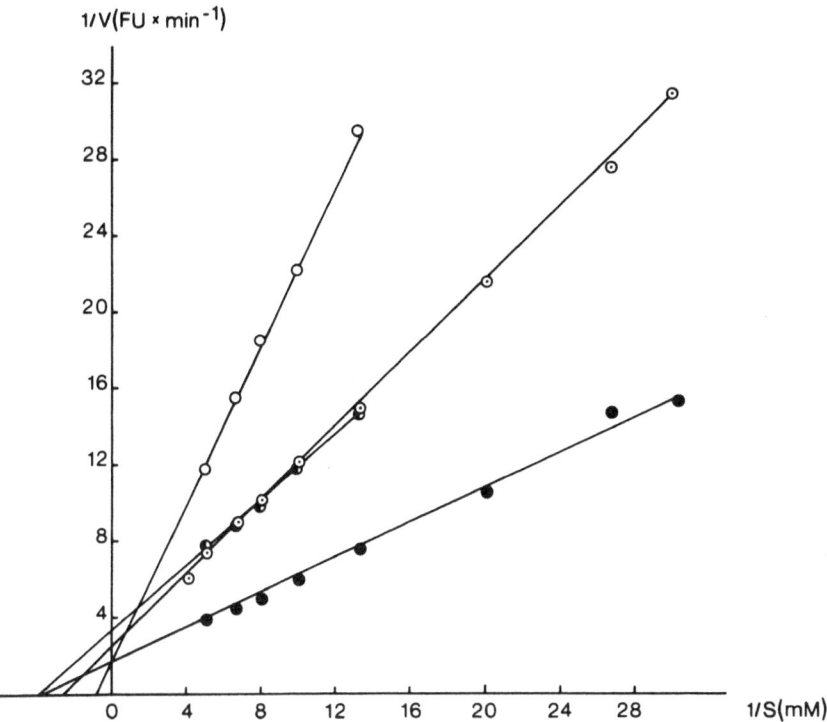

Fig. 36. APM activities plotted against various substrate concentrations ($\bullet$, $r = 0.985$) and angiotensins ($\circledcirc$ = 0.05 mM ANG I, $r = 0.978$; $\odot$ = 0.1 mM ANG II, $r = 0.998$; $\circ$ = 0.01 mM ANG III, $r = 0.993$) by the method of Lineweaver and Burk (1934). The APM activities were kinetically determined with the APM fluorometric medium in the ratio fluorometer, (specifications in Fig. 32; *FU*, fluorometric units) with 0.033–0.5 mM *L*-Ala-2NA in the renal homogenate (rat) (reaction temperature +25 °C). Each data point represents the mean of at least five series of measurements. The K_m is 0.24 mM. ANG III produces a competitive inhibition, and ANG I and II a noncompetitive inhibition

Microdissected glomeruli (Table 1): We were able to perform only a few biochemical fluorometric APA and APM measurements in microdissected glomerular preparations due to limited tissue quantities. According to these measurements, the K_m of glomerular APA is 0.23 mM. ANG I and II each produce a competitive type of inhibition. It is noteworthy that we found fairly high APM activities in our glomerular preparations, even though enzyme histochemical and immunohistochemical studies have previously indicated an absence of demonstrable APM there (Wachsmuth 1968, Wachsmuth and Torhorst 1974, Wachsmuth and Donner 1976, Kugler 1981b). We found a 1:6 ratio of APM to APA activities in our glomerular preparations. The explanation for this finding lies in the method of section preparation: Because the cryostat sections were "stacked" for freeze-drying, the glomeruli became contaminated with brush border fragments and their peptidases.

Besides biochemical fluorometric studies, we also performed *photometric APA measurements* in the renal homogenate of the rat. These were done essentially in connection with our quantitative histochemical work, the object being to determine the influence of the coupling agent FBB (purest-grade) and of various buffers and substrates on APA activities (Fig. 38).

33

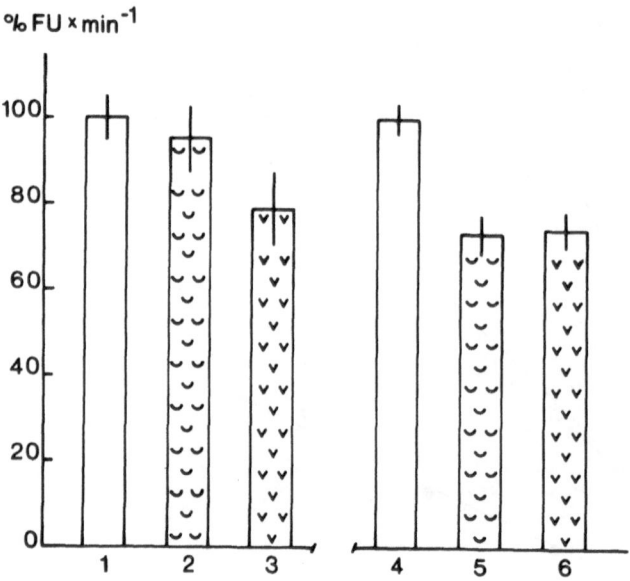

Fig. 37. Determination of APA (*columns 1–3*) and APM (*columns 4–6*) activities in the renal homogenate of the rat (biochemical fluorometric measurements) under the influence of the peptides *L*-Tyr-Ile (*columns 2 and 5*) and *L*-His-Pro-Phe (*columns 3 and 6*). The kinetic measurements of initial enzyme activities (*FU*, fluorometric units) were performed with the APA or APM fluorometric medium in the ratio fluorometer (specifications in Fig. 32; reaction temperature +25 °C). At least five measurements were performed for each of the different incubation conditions (with and without 1.5 mM *L*-Tyr-Ile or *L*-His-Pro-Phe). The APA activities of columns 2 and 3 were set in relation to column 1 (= 100%), and those of columns 5 and 6 to column 4 (= 100%). Description of results in text

In photometric investigations with the azo coupling method, it is crucial that specified times be adhered to in all stages of the procedure, since azo dye forms spontaneously in the medium, and the stability of the azo dye is very low after termination of the reaction. Even in the blank measurements (without substrate), these specified times must be maintained for all steps of the investigation. Having taken these precautions, we found that APA activities decline steadily with increasing concentrations of purest-grade FBB (from 0.25 to 1 mg/ml) (Fig. 38), showing a sizable scatter in some cases. Spontaneous azo dye formation is inversely proportional to the APA activities; i.e., as FBB concentrations are increased, APA activities decrease and spontaneous azo formation increases. This means that with 0.25 mg/ml FBB, spontaneous azo dye formation accounts for about 7% of the total measurable activity, and with 1 mg/ml, more than 10%. With the concentration of 0.5 mg/ml FBB used in our qualitative and quantitative histochemical studies, we calculate about a 5% mean inhibition of APA activities (corrected for spontaneous azo dye formation).

Our photometric measurements with various buffers and substrates show that the reaction rates of α-L-Glu-2NA and α-L-Glu-MNA are higher in cacodylate buffer than in tris-maleate buffer (Fig. 38; cf. Sect. 3.1.3). Comparing the two substrates, we find that the unsubstituted α-L-Glu-2NA leads to higher overall APA activities, especially in cacodylate buffer.

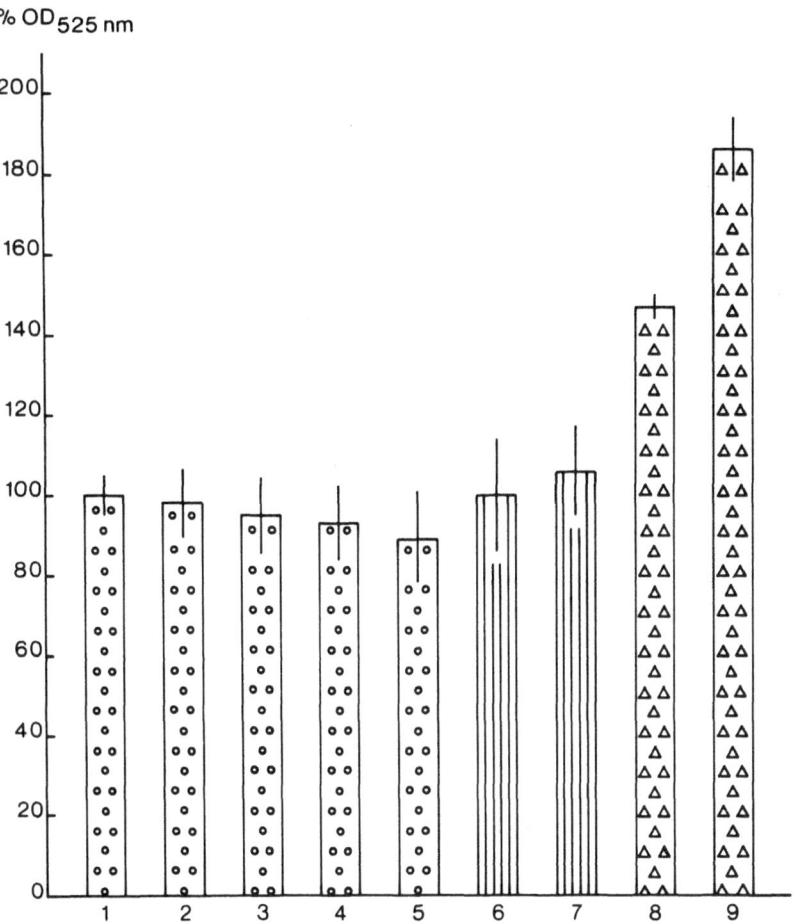

Fig. 38. Photometric APA measurements in the renal homogenate of the rat. The measurements were performed with the APA standard medium for qualitative histochemistry (simultaneous azo coupling) in semimicrocuvettes (light path 1 cm) at 525 nm and a reaction temperature of +22 °C. The effect of various concentrations of purest-grade FBB (*columns 1–5*) on APA activities was determined by means of endpoint measurements (*column 1* = without FBB; *2* = 0.25 mg/ml; *3* = 0.50 mg/ml; *4* = 0.75 mg/ml; *5* = 1.00 mg/ml). The reaction was run for 3 min and then stopped with 2 ml distilled water (0 °C), 100 μl of 1 N HCl was added 1.5 min later, and the measurements were performed after another minute had passed. For incubation without FBB (column 1), purest-grade FBB (1 mg/ml) was added only after the reaction, along with the 2 ml distilled water (post-coupling). Columns 2–5 were set in relation to column 1 (= 100%).

The enzymic attack of α-L-Glu-MNA and α-L-Glu-2NA was determined in tris-maleate and cacodylate buffer (*columns 6–9*) on the basis of kinetic measurements (simultaneous azo coupling). *Column 6* = α-L-Glu-MNA + tris-maleate buffer; *7* = α-L-Glu-MNA + 0.1 *M* cacodylate buffer; *8* = α-L-Glu-2NA + tris-maleate buffer; *9* = α-L-Glu-2NA + 0.1 *M* cacodylate buffer. The APA activities of columns 7–9 were set in relation to that of column 6 (= 100%). Description of results in text

3.2 Histochemical Localization of APA and APM in the Normal Kidney (Rat and Mouse)

APA and APM in the kidney are localized mainly in three different regions of the nephron: APA in the juxtaglomerular apparatus, renal corpuscle, and proximal tubule (Figs. 39, 40), and APM in the proximal tubule (Figs. 41, 42). Except where otherwise indicated, the APA and APM standard media were used on acetone-pretreated cryostat sections for light-microscopic localization of the enzymes.

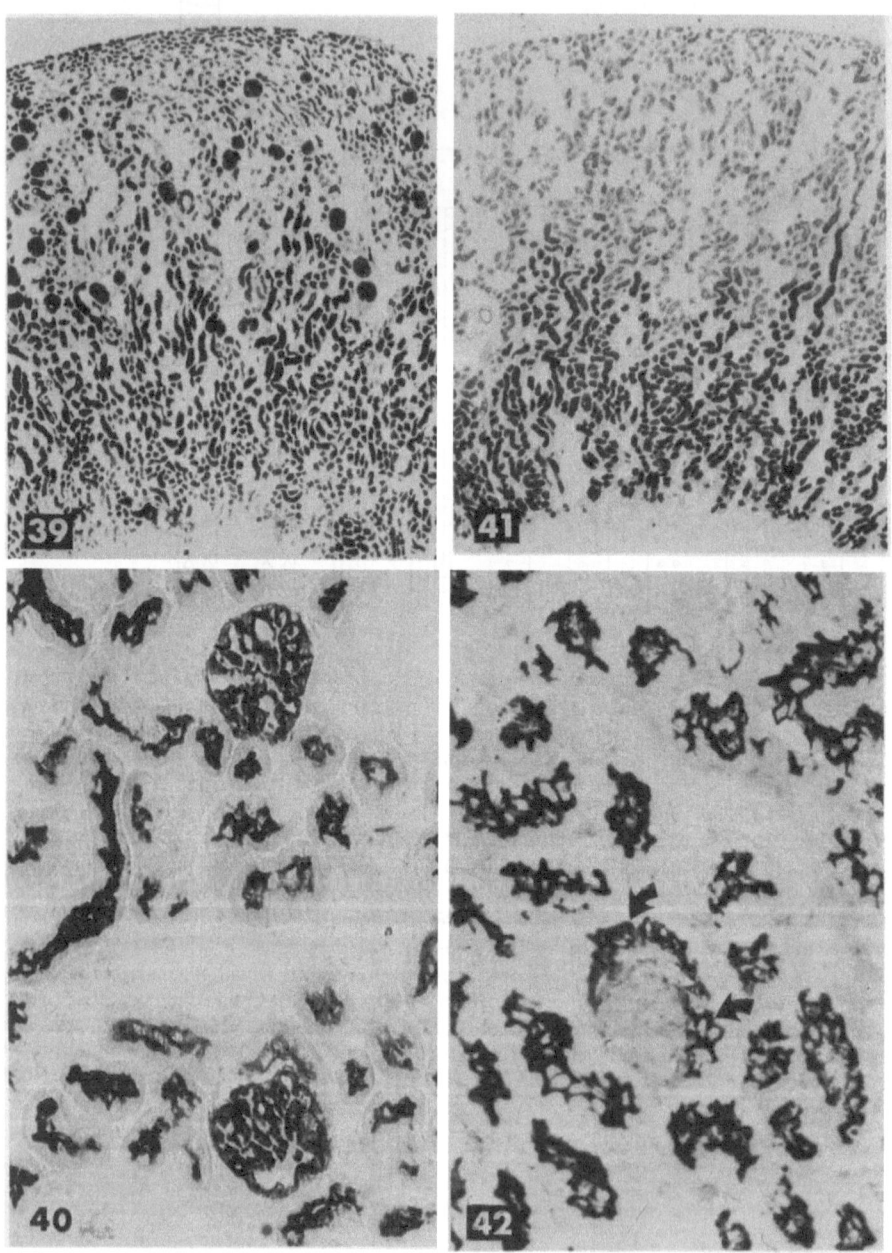

3.2.1 Juxtaglomerula Apparatus (JGA)

The JGA, located at the vascular pole of the renal corpuscle, is comprised of the following closely adjacent structures (cf. Bucher and Kaissling 1973, Gorgas 1978, Bargmann 1978 lit.): (a) granular and nongranular epithelioid cells in the pre- and postglomerular arterioles; (b) Goormaghtigh's cells, which form a nest in the angle between the afferent and efferent arteriole (Goormaghtigh 1932, Zimmermann 1933); and (c) the macula densa, a modified epithelial zone in the distal tubule (Zimmermann 1933).

Of the two enzymes investigated, only APA is histochemically demonstrable in the JGA. Different structures of the JGA react during the demonstration of APA in the rat and mouse. Common to both species, however, is the absence of APA in the macula densa.

Goormaghtigh's Cells. A positive APA reaction in these cells is observed only in rats (Fig. 43). Two findings are significant: (a) the Goormaghtigh's cells directly adjacent to the macula densa react more intensely than those bordering the mesangium, and (b) the activity of the APA demonstrable in Goormaghtigh's cell nests varies greatly from one JGA to the next and ranges from negative to strongly positive. Positive nests of Goormaghtigh's cells are generally more common in subcapsular and intermediate nephrons than in juxtamedullary nephrons and are observed more frequently in estrous female rats than in males.

Besides the Goormaghtigh's cells, certain vascular segments located near the glomerulus also show a reaction (Fig. 44), though no clear differentiation can be made between the afferent and efferent vessel.

Ultracytochemically, APA is demonstrated mainly at Goormaghtigh's cell membranes (Fig. 45), especially in zones of cell contact.

Epithelioid Cells. These cells are positive for APA only in the mouse (Figs. 46, 47). Not all epithelioid cells and cell complexes react uniformly in these animals. The activities vary from slightly positive to strongly positive. Ultracytochemically, the reaction products are localized mainly at epithelioid cell membranes (especially cell contacts) as well as in lysosomal structures to some degree (cf. Sect. 3.3). The cell membrane reactions are more distinct at a reaction-medium pH of 6.5 (or 7) than at pH 5. No sex differences are apparent.

Fig. 39. Male rat. Demonstration of APA. Incubation of 10-μm acetone-pretreated section in APA standard medium for 20 min at +30 °C. A reaction occurs in the glomeruli and brush borders of the proximal tubule. ×16

Fig. 40. Female mouse. Demonstration of APA. Specimen preparation and localization of APA as in Fig. 39. ×180

Fig. 41. Male rat. Demonstration of APM. Incubation of 10-μm acetone-pretreated section in APM standard medium for 5 min at +30 °C. The enzyme is localized in the brush borders of the proximal tubule, especially in the subcortical zone. ×45

Fig. 42. Female mouse. Demonstration of APM. Specimen preparation as in Fig. 41. The brush borders of the proximal tubule and the "tubule-like cells" of Bowman's capsule (*arrows*) are APM-positive. No reaction in the glomeruli. ×180

3.2.2 Renal Corpuscle

This consists of Bowman's capsule, the glomerular capillaries and their podocytes, and the highly cellular connective-tissue mesangium (Bargmann 1978 lit.). Bowman's capsule shows the following morphologic differences between adult mice and rats that have a bearing on the histochemical demonstration of enzymes: While in the rat Bowman's capsule has a monolayer squamous epithelium, many Bowman's capsules

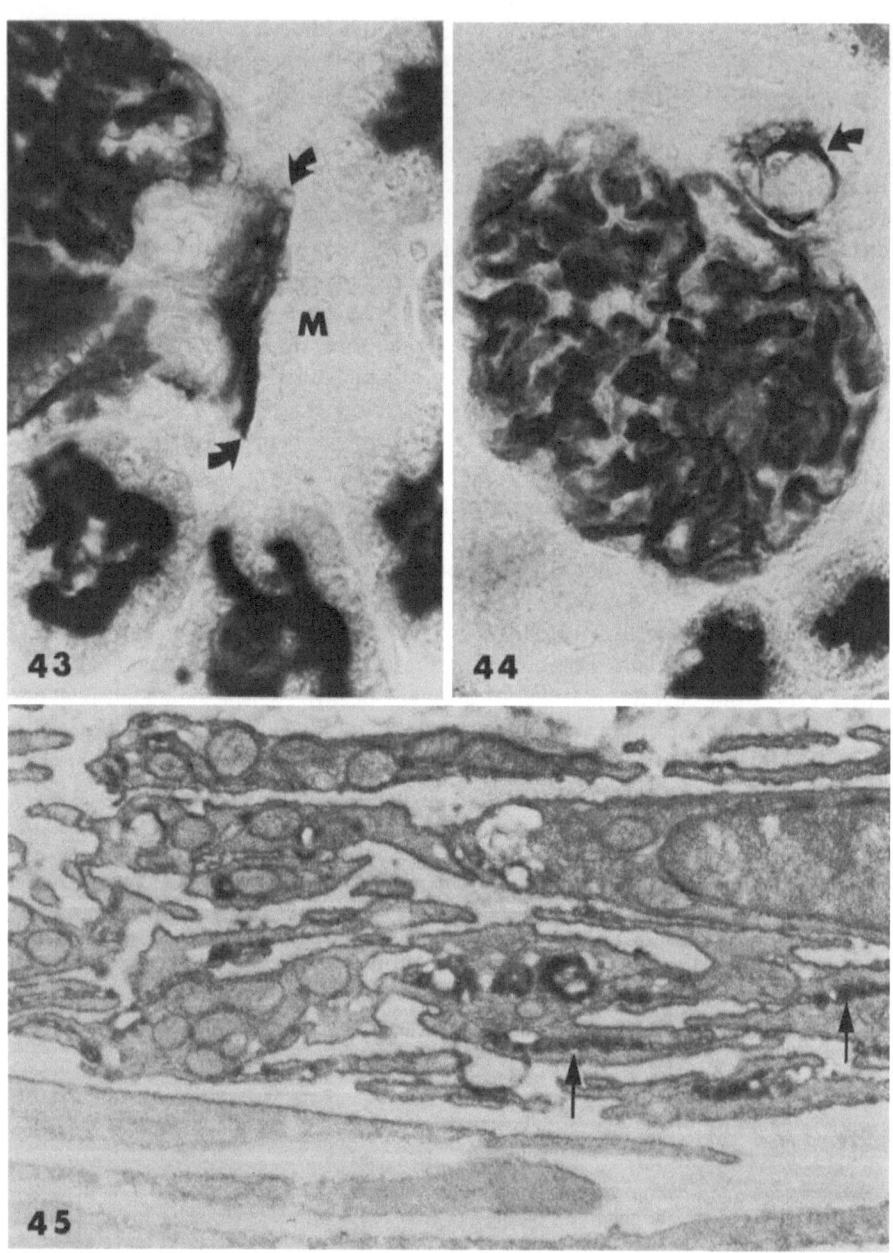

in the mouse possess isoprismatic to highly prismatic epithelial cells with brush borders ("tubule-like cells"; Von Möllendorf 1930, Helmholz 1935, Hanker et al. 1975; Kugler 1981b). According to Crabtree (1940), 94% of the renal corpuscles in adult male mice and 50% of those in adult females contain an isoprismatic parietal epithelium.

In the *rat*, only APA is demonstrable in the renal corpuscle (Figs. 43, 44), where its activity is about 31% of that in the strongly positive S_3 segment (Fig. 31) (male rat). The positive components of the renal corpuscle are the podocytes and endothelial cells. The epithelial cells of Bowman's capsule are negative for APA. Ultracytochemically, the reaction products are localized at the membranes of the aforementioned cell types.

In the glomerulus of the *mouse*, as in the rat, only APA (Figs. 46, 47) is demonstrable in podocytes (Figs. 48, 49) and endothelial cells (Fig. 50). Besides the membrane reactions, there is a conspicuous formation of reaction product in the area of the nuclear membrane of the podocytes (Figs. 51–53). The tubule-like cells of Bowman's capsule that occur predominantly in male mice contain both APA (Fig. 54) and APM (Fig. 55) in their brush borders.

An APA reaction is not clearly observed in mesangial cells of the rat and mouse.

The quantitative histochemical measurements in the rat and mouse glomerulus indicate that the mouse glomeruli contain about 50% higher APA activities.

3.2.3 Proximal Tubule

The proximal tubule of the rat and mouse possesses a brush border in all three segments (classification of Maunsbach 1966, cf. Bargmann 1978 lit.). APA and APM are demonstrable mainly in these brush borders in the rat and mouse.

In the proximal tubule of the *male rat*, the highest APA activities are localized in the S_3 segment (subcortical zone) (Figs. 56, 58). Taking as 100% the APA activities in the S_3 segment of male rats as determined by quantitative histochemical measurements, the APA activities in S_2 are approximately 63%, and in S_1 approximately 14% (Fig. 31). Accordingly, the lowest APA activities in the proximal tubule occur in the S_1 segment (Fig. 56).

There are additional reaction sites in the proximal tubule besides the brush borders; these are the vesicular structures (Fig. 57). These sites are best demonstrated in freeze-dried cryostat sections at pH 7 and at pH 5. These APA-positive vesicles

---

Fig. 43. Male rat. Demonstration of APA. Incubation of a 10-μm acetone-pretreated section in APA standard medium for 20 min at +30 °C. A strongly APA-positive Goormaghtigh's cell nest (*arrows*) is observed at the vascular pole of the glomerulus. It is located between the sectioned afferent and efferent vessels of the glomerulus and the macula densa (*M*). ×560

Fig. 44. Male rat. Demonstration of APA. Specimen preparation as in Fig. 43. Next to the glomerulus is a sectioned vessel (*arrow*) with APA-positive cells. ×560

Fig. 45. Male rat. Ultracytochemical demonstration of APA in the Goormaghtigh's cells. The enzyme demonstration was performed by simultaneous azo coupling (α-L-Glu-MNA as substrate, HPR as coupling agent) at pH 5 by 120-min incubation at room temperature. The reaction product is localized at cell membranes (*dark lines*) and especially in the region of cell contacts (*arrows*). The light spaces between the JG cells or cellular extensions contain material from basement membranes, which are poorly visualized with this technique (unstained). ×21 000

occur mainly in the S_2 segment and the beginning of S_3. They have no preferential localization but are distributed uniformly throughout the cell. Both the vesicular and brush-border reactions show sex differences in the rat. Estrous females show a greater abundance of APA-positive vesicles in the aforementioned segments, although the brush borders in the terminal portions of S_3 are nearly negative (Fig. 59). By contrast, this terminal portion of the S_3 segment is strongly positive in males (cf. Fig. 58, quantitative distribution of APA).

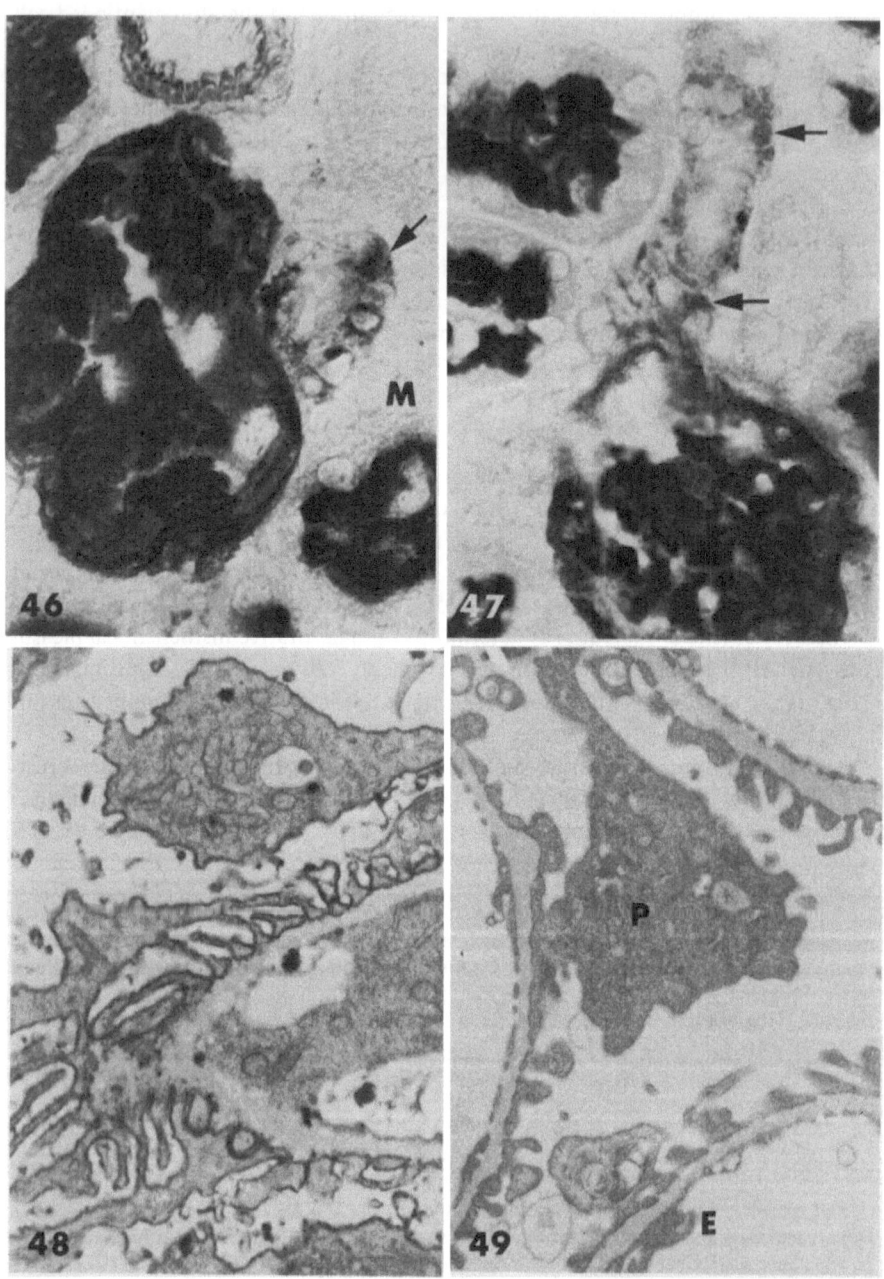

An increase in APM activity from S_1 to S_3 in the brush border of the rat proximal tubule is also demonstrable by means of qualitative histochemistry (Fig. 62). This increase is especially marked in male rats, whose subcortical zone shows a stronger overall reaction than in females (Fig. 62). We found no intracellular reaction sites for APM.

In the *mouse*, no sex differences are apparent on the demonstration of APM and APA in the brush border of the proximal tubule. There are differences in the localization of these two enzymes, however. APA is observed in the brush borders of the S_1 and S_2 segments, but not in S_3 (Fig. 60). No activity differences between the S_1 and S_2 segments are apparent on qualitative histochemical study. On the intracellular level there are some clump-like structures that are positive for APA and, especially in FDC sections (pH 7), supranuclear vesicular APA-positive structures that resemble Golgi elements.

In contrast to APA, APM is present in all three segments of the proximal tubule in the mouse (Fig. 61), the highest activities occurring in the brush border of the S_3 segment. No intracellular reaction sites are observed for APM.

3.2.4 Additional Reaction Sites of APA and APM

APA is demonstrable in the following additional renal structures: (a) cell membranes of capillary endothelia of the renal medulla of the rat and mouse (Figs. 63, 64; this reaction is very weak in the rat); (b) peritubular capillaries (rarely); (c) the vasa recta of the renal medulla (markedly; Figs. 63, 65) (cf. Lojda and Gossrau 1980) as well as in the muscle cell membranes of the vasa arcuata, interlobularia, afferentia, and efferentia in the mouse (Fig. 66) (Kugler 1981b); (d) the descending limbs of the loops of Henle in the mouse (APA activities here are very slight and may not occur at all).

Besides the brush borders, *APM* is demonstrable in a certain cell population of the connecting tubules and collecting ducts; this localization is more prominent in the rat than in the mouse (Fig. 67). The reaction is these cells is confined to apical regions and is found in cortical as well as medullary (down to and including the inner stripe of the outer zone) segments of the collecting ducts (cf. Kugler 1981b).

Fig. 46. Female mouse. Demonstration of APA. Incubation of 10-μm acetone-pretreated section in APA standard medium for 20 min at +30 °C. Besides the glomerulus the epithelioid cells (*arrow*) of the afferent arteriole (polkissen) contain reaction products (partly in granular form). *M*, macula densa. ×560

Fig. 47. Female mouse. Demonstration of APA. Specimen preparation as in Fig. 46. An afferent arteriole with APA-positive epithelioid cells (*arrows*) passes to the glomerulus. ×560

Fig. 48. Female mouse. Ultracytochemical demonstration of APA in the podocytes of the glomerulus. The demonstration was done with α-L-Glu-MNA as substrate and HPR as coupling agent at pH 6.5 (incubation 120 min at room temperature). Reaction product is found at the cell membranes of the podocytes. ×17 500

Fig. 49. Female mouse. Ultracytochemical control reaction for APA demonstration in podocytes and endothelial cells. Specimen preparation as in Fig. 48, but without α-L-Glu-MNA in the incubation medium. No reaction in podocytes (*P*) or endothelial cells (*E*) (cf. Figs. 48, 50). ×17 500

3.3 Animal Experiments

The changes in body weight and adrenal weight of male rats associated with various dietary regimens and adrenalectomy are shown in Table 2. It is seen that the adrenal weights in animals fed a high-sodium diet are 13% lower on average than in control

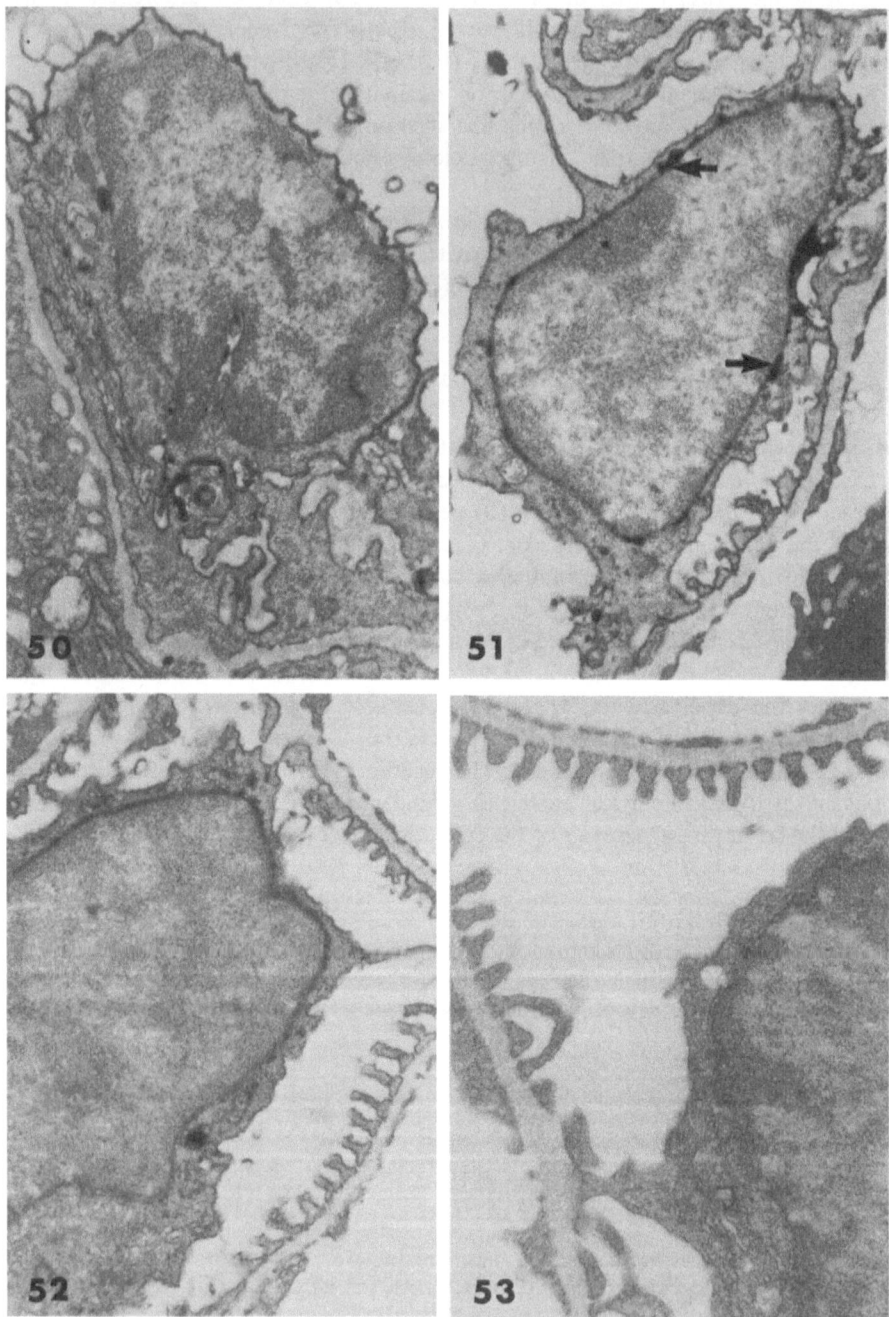

animals; that the adrenal weights of sham-operated animals are 10% higher than in normal animals; and that adrenalectomy is followed by a massive weight loss during the observation period (9 days). However, the differences between adrenal weights are not statistically significant by the U test (95% level). All three diets produce roughly the same weight gain over the observation period, although rats maintained on a standard diet during the same period show an approximately 20% smaller weight gain.

3.3.1 Low-Sodium Diet

Histochemical Investigations

Juxtaglomerular Apparatus

The number of APA-positive *Goormaghtigh's cell nests* in the rat appears to be slightly higher relative to control animals (Altromin C 1000) (Fig. 69). The distribution pattern of positive Goormaghtigh's cell nests within the renal cortex appears to be unchanged relative to normal animals, however (APA-positive Goormaghtigh's cell nests are mainly subcapsular).

In mice there is a slight increase in APA-positive *epithelioid cells*, while the APA reaction in muscle cell membranes of intrarenal arterial vessels appears to be diminished.

Renal Corpuscle

Qualitative histochemistry reveals an increased APA activity in the glomeruli of the rat (Fig. 68) and mouse. This is confirmed by quantitative histochemical work in subcapsular glomeruli (Fig. 86), which shows an APA activity increase of about 40% in rats and about 31% in mice. These changes are statistically significant by the U test (99% level, Table 3) in both rats and mice compared to control animals (Altromin C 1000).

Fig. 50. Male mouse. Ultracytochemical demonstration of APA in endothelial cells of the glomerulus. Specimen preparation as in Fig. 48. Reaction product is found at the cell membranes (*dark lines*). ×15 700

Fig. 51. Male mouse. Ultracytochemical demonstration of APA in podocytes of the glomerulus. Specimen preparation as in Fig. 48. Besides the cell membrane, reaction product is observed in the region of the nuclear membrane (*arrows*). ×15 700

Fig. 52. Male mouse. Ultracytochemical demonstration of APA in podocytes of the glomerulus. The demonstration was performed with α-L-Glu-MNA as substrate and HPR as coupling agent at pH 5 (incubation 120 min at room temperature). The reaction sites at pH 5 correspond to those at pH 6.5 (cf. Fig. 51), but there is less reaction product in the region of the nuclear and cell membranes. ×15 700

Fig. 53. Male mouse. Ultracytochemical control reaction for APA demonstration in podocytes and endothelial cells. Specimen preparation as in Fig. 52, but without the substrate α-L-Glu-MNA in the incubation medium. No reaction product in podocytes or endothelial cells (cf. Fig. 52). ×19 200

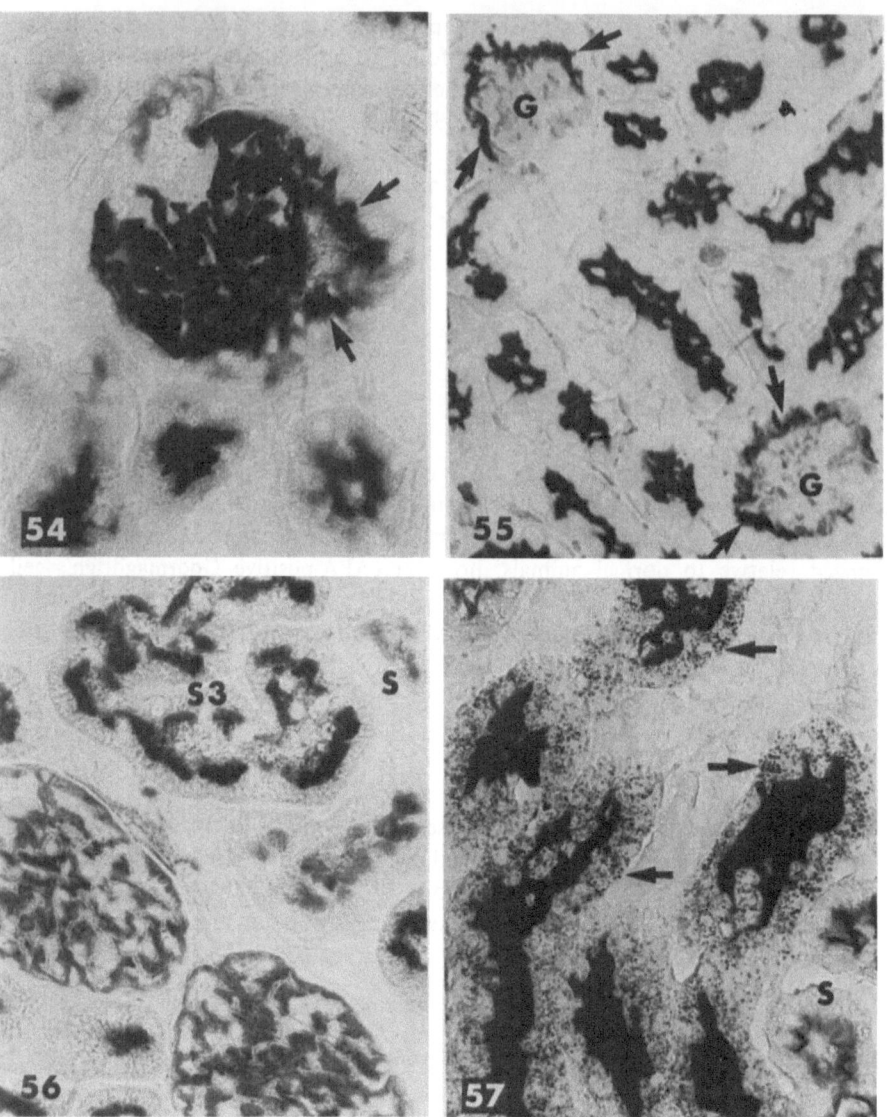

Fig. 54. Male mouse. Demonstration of APA. Incubation of 10-μm acetone-pretreated cryostat section in APA standard medium for 20 min at +30 °C. A strong APA reaction is found in the glomerulus and tubule-like cells of Bowman's capsule (*arrows*), and a weak reaction in epithelioid cells of the afferent arteriole. ×250

Fig. 55. Male mouse. Demonstration of APM. Incubation of 10-μm FDC section in APM standard medium for 60 min at +30 °C. Besides the brush borders of the proximal tubule, the tubule-like cells (*arrows*) contain high APM activities. The glomeruli (*G*) are APM-negative. ×240

Fig. 56. Male rat. Demonstration of APA. Specimen preparation as in Fig. 54, but with a 4-μm-thick section. The thinner section provides a sharp localization of reaction product. High activities are demonstrated in the S_3 segment (S_3), moderate activities in the glomeruli, and low activities in the S_1 segment of the proximal tubule (*S*). ×350

Fig. 57. Female rat. Demonstration of APA. Incubation of 10-μm FDC section in APA standard medium for 60 min at +30 °C. Besides the brush-border reactions, a number of APA-positive intracellular vesicles (*arrows*) are observed in the S_2 segments. None are found in the S_1 segment (*S*) of the proximal tubule. ×250

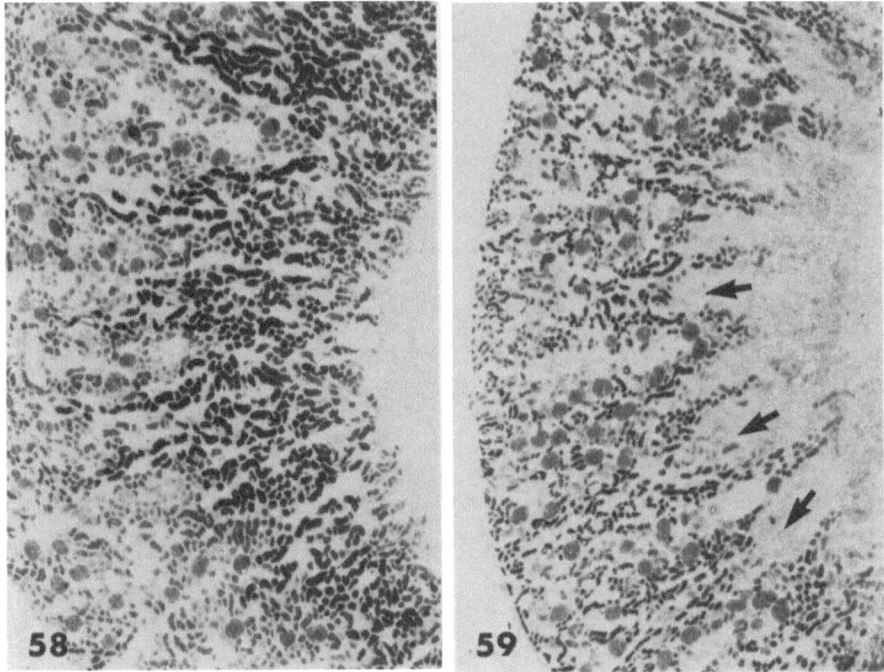

Fig. 58. Male rat. Demonstration of APA. Incubation of 10-μm acetone-pretreated section in APA standard medium for 20 min at +30 °C. Besides the glomeruli, the brush borders of the proximal tubule display a positive APA reaction. APA activities are particularly high in the S_3 segment of the proximal tubule, which results in a stronger APA reaction in the subcortical zone of male rats. ×16

Fig. 59. Female rat. Demonstration of APA. Specimen preparation as in Fig. 58. In contrast to the male rat (cf. Fig. 58), the terminal portions of the S_3 segments in the female rat show no APA activity (*arrows*), with on otherwise identical reaction pattern; this results in a weaker APA reaction in the subcortical zone of females. ×18

Proximal Tubule

No qualitative histochemical differences are apparent between treated and control animals.

Biochemical Investigations

The fluorometric measurements of renal homogenate were performed for both APA and APM (Figs. 87, 88). Measurements in the control animals (Altromin C 1000) indicate that APA activities in the rat kidney are five times higher than in the mouse kidney, and that APM activities are more than seven times higher, based on specific activity values (mU/mg protein). In rats fed a low-sodium diet we find a decline of about 10% in mean APA activities and about 14% in mean APM activities. Mice fed the same diet show no substantial changes in APA activities (about +7%) or APM activities (about −2%). The U test, however, indicates that neither species shows a statistically significant response (95% level) to low-sodium feeding in terms of APA and APM activity changes relative to the controls (Altromin C 1000, Table 3).

45

3.3.2 High-Sodium Diet

Histochemical Investigations

Juxtaglomerular Apparatus

Compared with controls, rats fed a high-sodium diet show a slight decrease in the number of APA-positive *Goormaghtighs's cell nests*. In mice we find less reaction

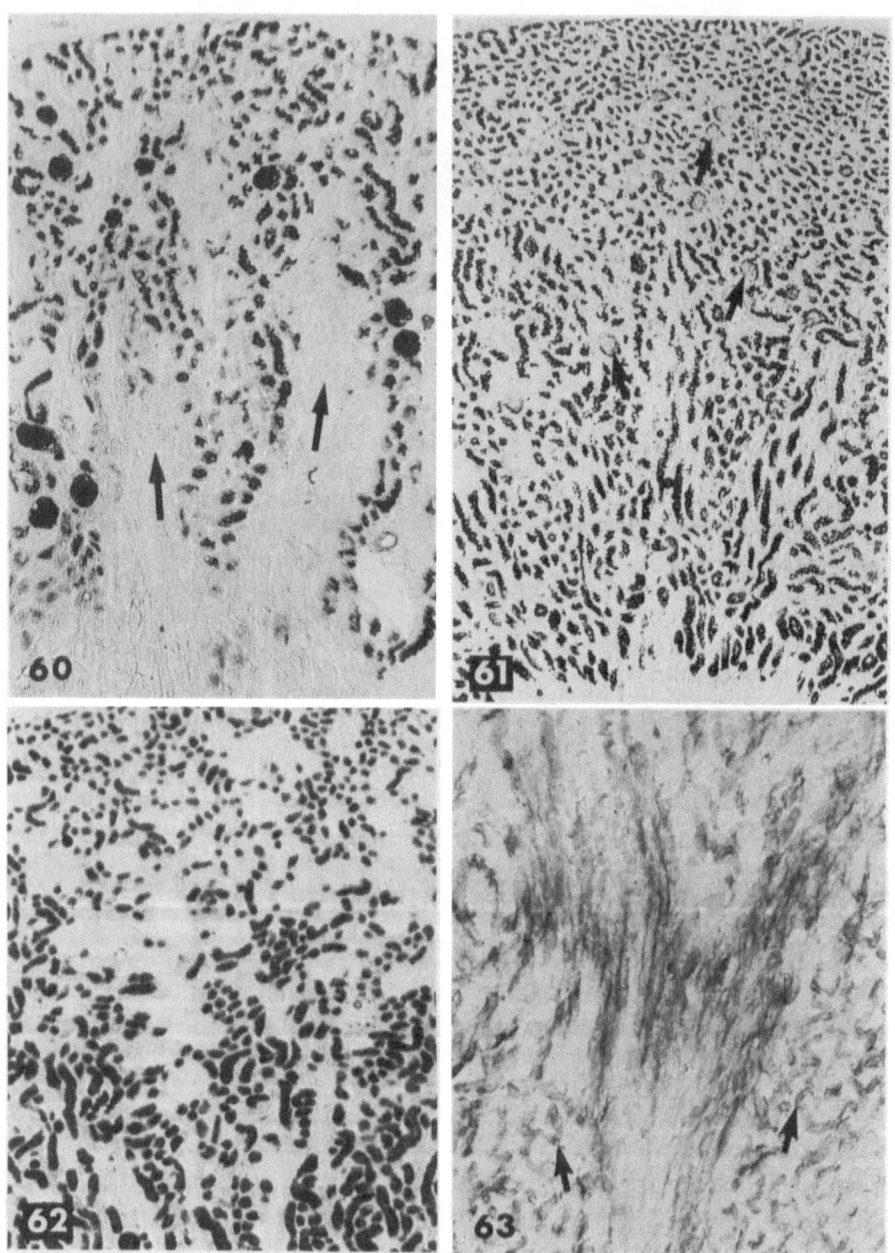

Table 2. Changes in the body weights and adrenal weights of experimental animals (five to ten male rats per experiment)

Treatment and treatment period[a]	Change in body weight during treatment period	Wet weight of adrenals per 100 g BW
Normal animals-TPF 1320 and water, 20 days	Gain of 33.3 g ± 7.4 g	17.5 mg ± 2.6 mg
Sham-operated-TPF 1320 and water 9 days	Gain of 15.0 g ± 2.8 g	19.3 mg ± 2.0 mg
Bilateral adrenal-ectomy-TPF 1320 and water, 9 days	Loss of 55.0 g ± 8.4 g	–
Control diet C 1000 and bidist. water, 20 days	Gain of 41.6 g ± 8.8 g	16.0 mg ± 3.2 mg
Low-Na diet C 1036 and bidist. water, 20 days	Gain of 36.7 g ± 12.7 g	15.6 mg ± 2.4 mg
Control diet C 1000 and 1% NaCl in bidist. water, 20 days	Gain of 41.5 g ± 13.0 g	13.9 mg ± 1.0 mg

[a]The diets were obtained from the Altromin Co. The animals received food and liquids ad libitum; stall temperature $21° ± 2 °C$.

Fig. 60. Male mouse. Demonstration of APA. Incubation of 10-μm acetone-pretreated section in APA standard medium for 20 min at +30 °C. High APA activities are observed in glomeruli and brush borders of the S_1 and S_2 segments of the proximal tubule. The S_3 segments (*arrows*) are APA-negative. ×35

Fig. 61. Male mouse. Demonstration of APM. Incubation of 10-μm FDC section in APM standard medium for 60 min at +30 °C. Unlike APA (cf. Fig. 60), APM is not present in glomeruli (*arrows*), but does occur in all proximal tubule segments, especially in the subcortical zone. ×35

Fig. 62. Female rat. Demonstration of APM. Incubation of 10-μm acetone-pretreated section in APM standard medium for 5 min at +30 °C. An APM reaction occurs only in the brush borders of all proximal tubule segments. Of these, S_3 develops the highest APM activities, though they are lower in the female rat than in the male. ×80

Fig. 63. Female mouse. Demonstration of APA in renal medulla. Specimen preparation as in Fig. 60. This section shows conspicuous APA activities in bundles of vasa recta and in capillaries (*arrows*). ×120

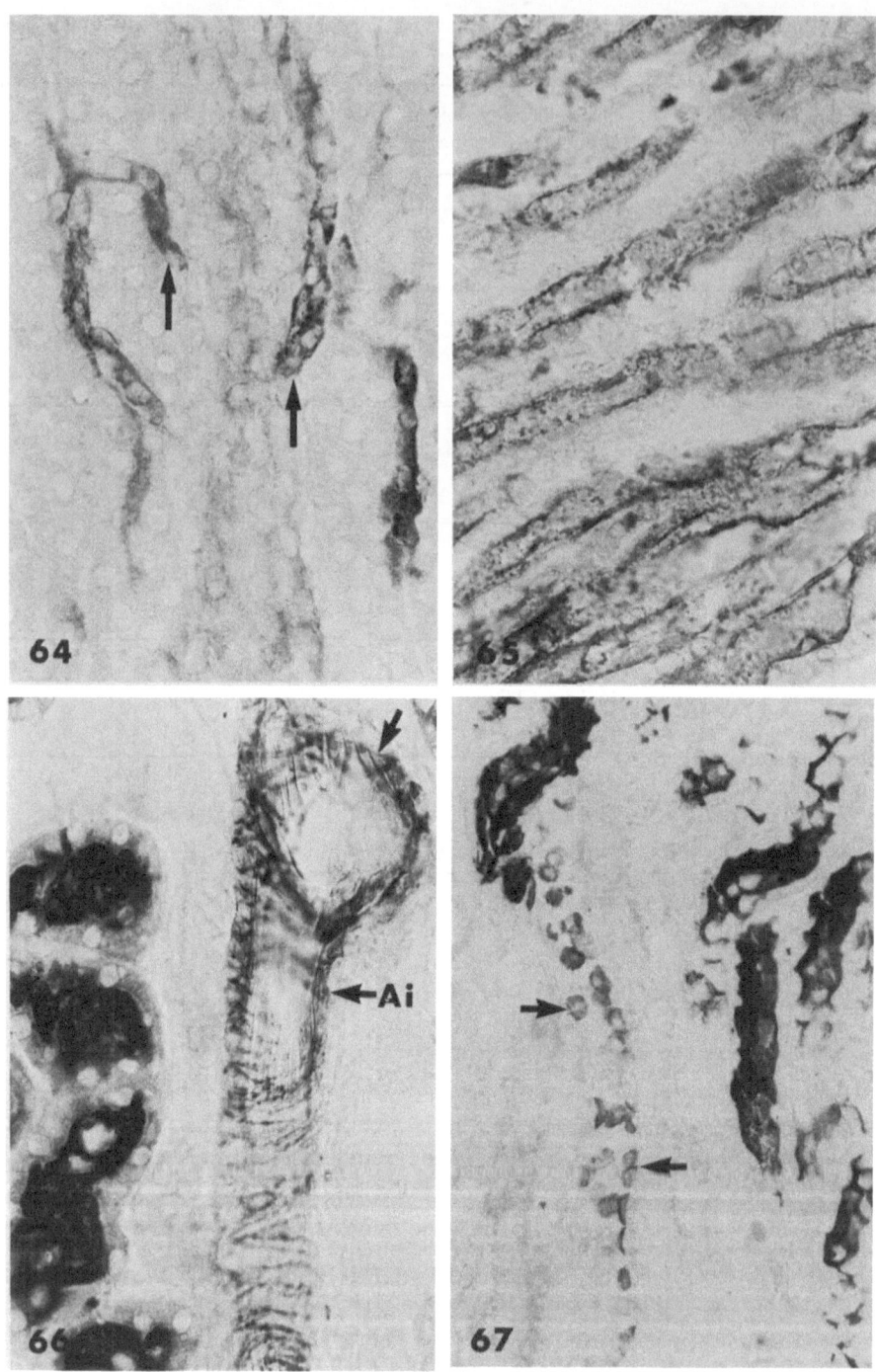

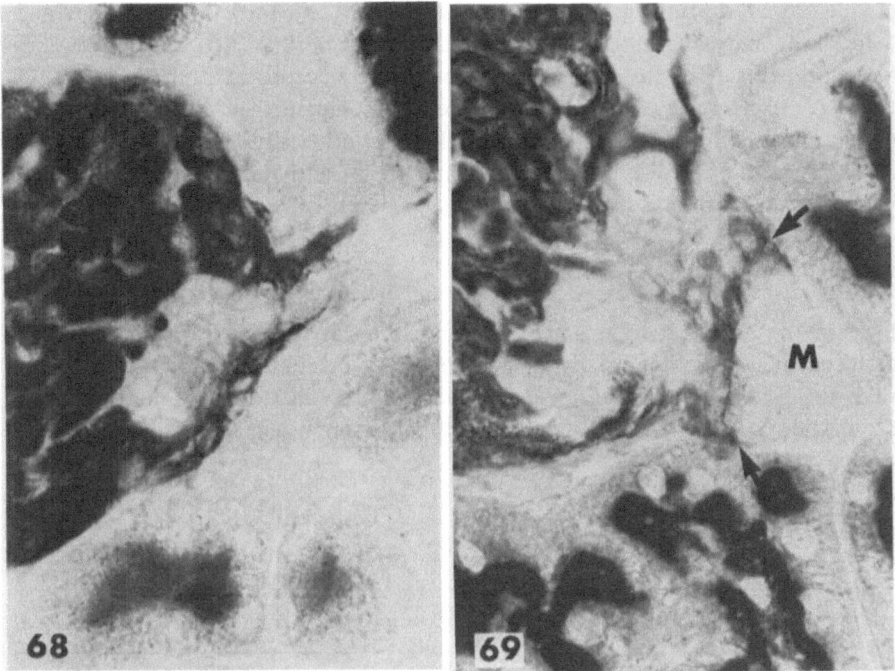

Fig. 68. Male rat fed a low-sodium diet (Altromin C 1036). Demonstration of APA. Incubation of 10-μm acetone-pretreated section in APA standard medium for 20 min at +30 °C. Besides the glomerulus, high APA activities are demonstrated in a Goormaghtigh's cell nest and an afferent or efferent vessel of the glomerulus. ×560

Fig. 69. Male rat (Altromin control diet C 1000). Demonstration of APA. Specimen preparation as in Fig. 68. This sectional plane reveals the nest-shaped arrangement of the APA-positive Goormaghtigh's cells (*arrows*) around the macula densa (*M*). This cell nest extends to the segments of the proximal tubule. ×560

product in the *epithelioid cells*, but more in the muscle cell membranes of intra-renal arteries.

Renal Corpuscle

Both qualitative and quantitative histochemical studies demonstrate a marked increase in the APA activities of the glomeruli compared with controls. Thus, accord-

---

Fig. 64. Enlarged section from Fig. 63 showing the capillary reaction (*arrows*) in the renal medulla. ×350

Fig. 65. Enlarged section from Fig. 63 showing APA activity in the vasa recta. ×350

Fig. 66. Male mouse. Demonstration of APA. Incubation of 10-μm section in APA standard medium for 20 min at +30 °C. The micrograph shows a cross-sectioned arcuate artery (*arrow*) from which an interlobular artery (Ai, *arrow*) arises. The muscle cell membranes of these vessels display an APA reaction. ×350

Fig. 67. Female rat. Demonstration of APM. Incubation of 10-μm FDC section in APM standard medium for 60 min at +30 °C. Besides the brush-border reaction in the terminal portions of S_3, there are marked reactions in individual cells of the collecting ducts (*arrows*). ×160

ing to densitometric findings, the APA activities are about 40% higher in rats and 26% higher in mice than in the controls (Altromin C 1000) (Fig. 86). These activity changes are statistically significant in both species according to the U test (99% level, Table 3). However, if we use the U test to compare the APA activities on a high-sodium and a low-sodium diet, we find that no statistically significant differences occur in the rat (95% level), but that mice show significantly higher APA activities on a low-sodium diet than on a high-sodium diet (99% level, Table 3).

Proximal Tubule

Changes in the APA and APM activities in brush borders of the proximal tubule of the rat and mouse are not clearly demonstrated by qualitative histochemical work. In mice the APA activities appear unchanged and the APM activities increased relative to the controls, while the APA and APM activities in rats appear to be diminished.

Table 3. Statistical results (U test) on APA and APM changes in the kidney of experimental animals

Quantitative histochemical APA determinations in glomeruli

		Rat			Mouse
Normal animal	>	Control animal[+]		–	
Low-sodium	>	Control animal[+]	Low-sodium	>	Control animal[+]
High-sodium	>	Control animal[+]	High-sodium	>	Control animal[+]
High-sodium	=	Low-sodium	Low-sodium	>	High-sodium[+]
Sham-operation	>	Normal animal[+]		–	
Adrenalectomy	>	Normal animal[+]	Mouse	>	Rat[+]
Adrenalectomy	>	Sham operation[+]		–	

Fluorometric APA- and APM-measurements in renal homogenate

		APA			APM
		Rat			Rat
Control animal	=	Low-sodium	Control animal	=	Low-sodium
Control animal	>	High-sodium[+]	Control animal	>	High-sodium[+]
Low-sodium	=	High-sodium	Low-sodium	=	High-sodium
Sham operation	>	Adrenalectomy[+]	Sham operation	=	Adrenalectomy
		Mouse			Mouse
Control animal	=	Low-sodium	Control animal	=	Low-sodium
Control animal	=	High-sodium	High-sodium	>	Control animal[+]
Low-sodium	=	High-sodium	High-sodium	>	Low-sodium[+]

+ Statistically significant at the 99% level
= Difference not statistically significant (95% level)

Biochemical Investigations

The fluorometrically measurable APA activities in renal homogenate are about 19% lower in rats and only about 3% lower in mice than in the controls (Figs. 87, 88). The decrease in rats is statistically significant by the U test (99% level), whereas the decrease in mice is not (95% level) (Table 3). A statistical comparison of the APA changes on a low- and high-sodium diet shows no significant differences (95% level, Table 3) in either species.

With regard to *APM* activities, marked differences are observed between the renal homogenates of the rat and mouse. While the activities in rats decrease by about 17%, those in mice show an increase of about 28% (Figs. 87, 88). In both species these changes are statistically significant (99% level in U test, Table 3) relative to the controls (Altromin C 1000). However, comparing the APM activities of animals fed high- and low-sodium diets, we find no significant difference in rats (95% level in U test), but significantly higher activities in mice fed a high-sodium diet (99% level, Table 3).

3.3.3 Adrenalectomy

Adrenalectomy leads to more pronounced histochemical and biochemical changes, compared with the dietary experiments.

Histochemical Investigations

Juxtaglomerular Apparatus

Goormaghtigh's Cells. An APA-positive reaction is demonstrated only in the Goormaghtigh's cells of the rat under these experimental conditions. There is a substantial increase in the number of APA-positive Goormaghtigh's cell nests, most of which show a very strong reaction (Fig. 70). Indeed, every Goormaghtigh's cell nest present in the renal section appears to show a reaction, thereby cancelling the numerical reaction gradient between the cortex and medulla. The Goormaghtigh's cell nests exhibit an appreciably greater surface area than in sham-operated controls, but even these controls show a noticeable increase in both the number and intensity of reacting Goormaghtigh's cell nests relative to normal animals (Fig. 71). It is also noteworthy that the reaction of the Goormaghtigh's cells is continued in the direction of the afferent and efferent arteriole. Again, ultracytochemical study localizes the APA to the membranes of the Goormaghtigh's cells.

Epithelioid Cells. The epithelioid cells show marked changes in the rat and mouse following adrenalectomy. Especially in *mice*, there is a strong hypertrophy of epithelioid cells or cell complexes, which demonstrate high APA activities on qualitative histochemical study (Figs. 72–77). These epithelioid cell reactions are most conspicuous in the afferent arteriole (Figs. 72–77), but are also present in the efferent arteriole to some degree (Figs. 73, 74). Individual cell reactions are also frequent in the interlobular artery. In light microscopy the reaction is localized mainly in the region of cell membranes and in intracellular vesicular structures. Ultracyto-

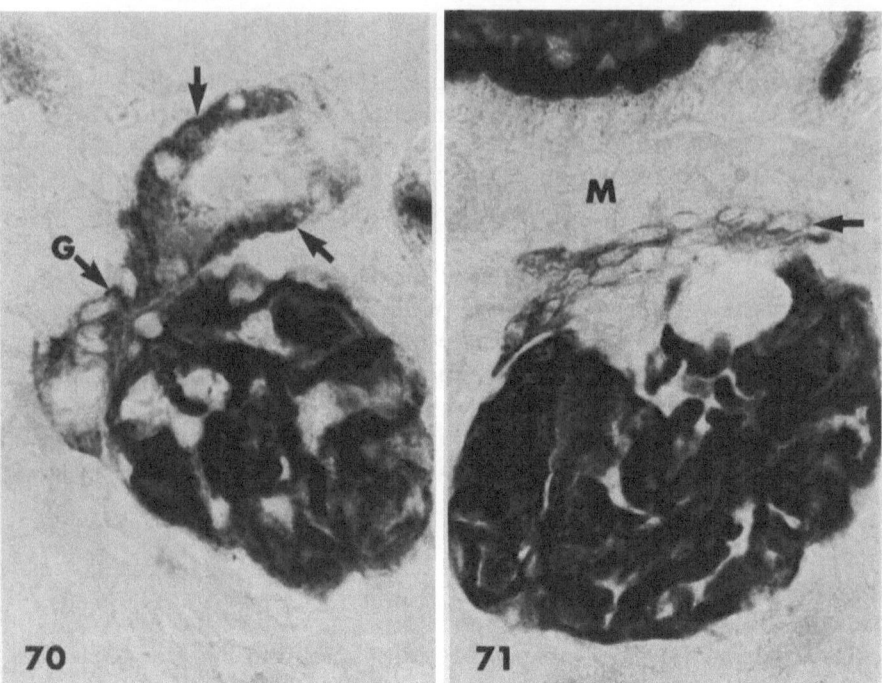

Fig. 70. Adrenalectomized male rat. Demonstration of APA. Specimen preparation as in Fig. 68. A Goormaghtigh's cell nest. (*G, arrow*) and a group of epithelioid cells (*arrows*) with high APA activities are found at the vascular pole of the strongly reacting, tangentially sectioned glomerulus. ×640

Fig. 71. Sham-operated male rat. Demonstration of APA. Specimen preparation as in Fig. 68. This sectional plane reveals a broad, flat APA-positive Goormaghtigh's cell nest (*arrow*) below the macula densa (*M*). The reaction-free Goormaghtigh's cell nuclei oriented tangentially to the glomerulus are clearly visible. ×560

Fig. 72. Adrenalectomized male mouse. Demonstration of APA. Incubation of 10-μm section in APA standard medium with HNF instead of FBB for 30 min at +30 °C. APA is observed at the vascular pole in a longitudinally sectioned afferent arteriole with a group of strongly positive epithelioid cells. Some reaction product is localized in vesicular structures. ×375

Fig. 73. Adrenalectomized male mouse. Demonstration of APA. Incubation of 10-μm acetone-pretreated section in APA standard medium for 20 min at +30 °C. Besides the strongly APA-positive glomerulus, high APA activities are demonstrated in hypertrophic epithelioid cells of the afferent arteriole (*arrow*) as well as in the efferent arteriole to some degree. *M*, macula densa. ×600

Fig. 74. Adrenalectomized male mouse. Demonstration of APA. Specimen preparation as in Fig. 73. Passing to the strongly APA-positive glomerulus is a tangentially sectioned afferent arteriole whose transversely oriented epithelioid cells display high APA activities (*arrows*). A cross-sectioned APA-positive efferent arteriole (*Ve*) is also seen at the vascular pole. ×375

Fig. 75. Adrenalectomized male mouse. Demonstration of APA. Specimen preparation as in Fig. 73. The efferent arteriole (*arrow*) of this glomerulus shows only low APA activities compared to the high activities in the afferent arteriole. ×375

chemical study further reveals that the membrane reactions occur primarily at points of cell contact (pH 7; Figs. 78—81), and that reaction products appear in lysosomal vacuolar and myelin-like laminar structures (especially at pH 5; Fig. 82).

In contrast to normal animals, diet animals, and sham-operated controls, APA reactions are demonstrable in epithelioid cells of *rats* after adrenalectomy (Figs. 83—85). Most of these occur in the afferent arteriole. Also noteworthy is the occurrence of epithelioid cell-like reactions of individual cells in the region of the interlobular artery (Fig. 84).

Renal Corpuscle

Qualitative study shows a marked activity increase in the glomeruli of the rat and mouse relative to sham-operated controls. Quantitative histochemistry reveals that

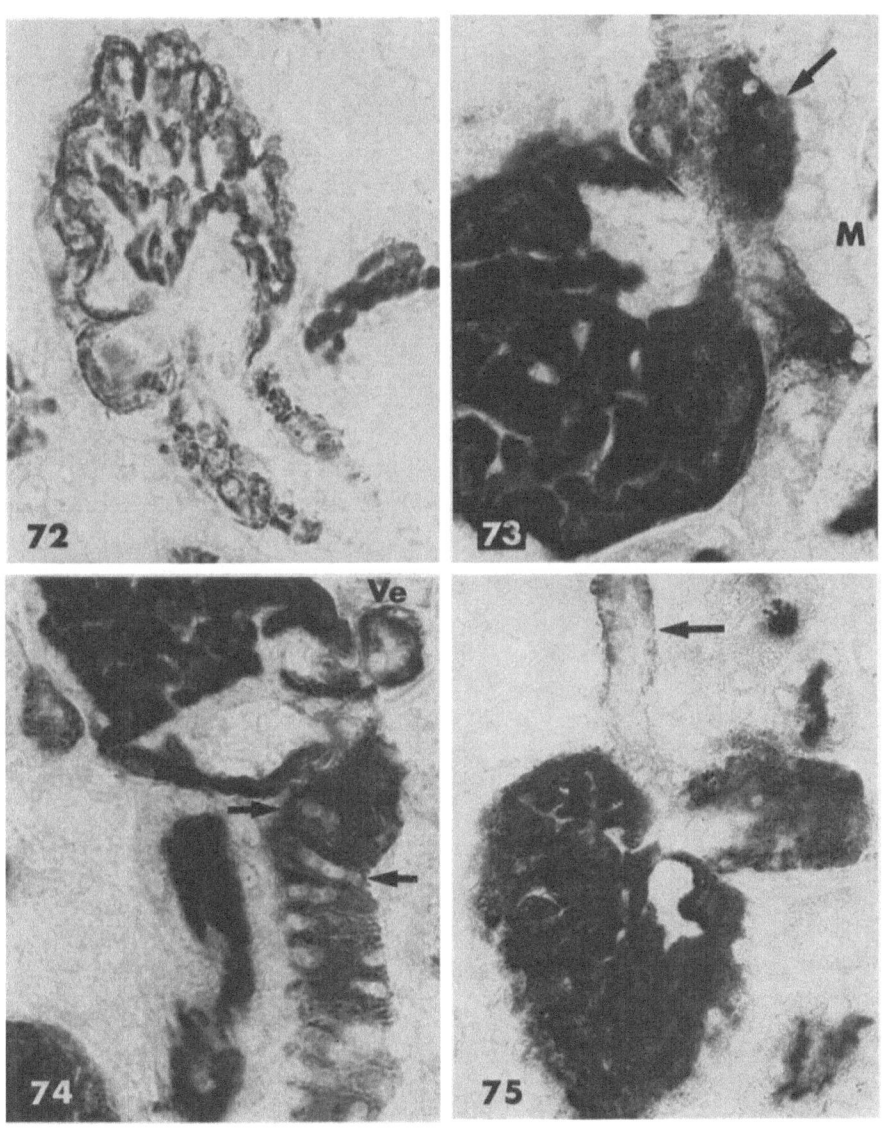

this increase is approximately 28% in rats compared with sham-operated animals and approximately 35% compared with normal animals (Fig. 86). This shows that the sham operation itself produces a slight increase in APA activities (about 6%) relative to normal animals. It is important to note that these changes in shamoperated animals relative to normal animals and in adrenalectomized animals relative to sham-operated animals are statistically significant by the U test (99% level, Table 3).

Also noteworthy is the reduction in the brush borders of tubule-like cells of Bowman's capsule in adrenalectomized mice, as well as the frequent flattening of this epithelium. These changes are accompanied by a parallel decrease in demonstrable APA and APM activities.

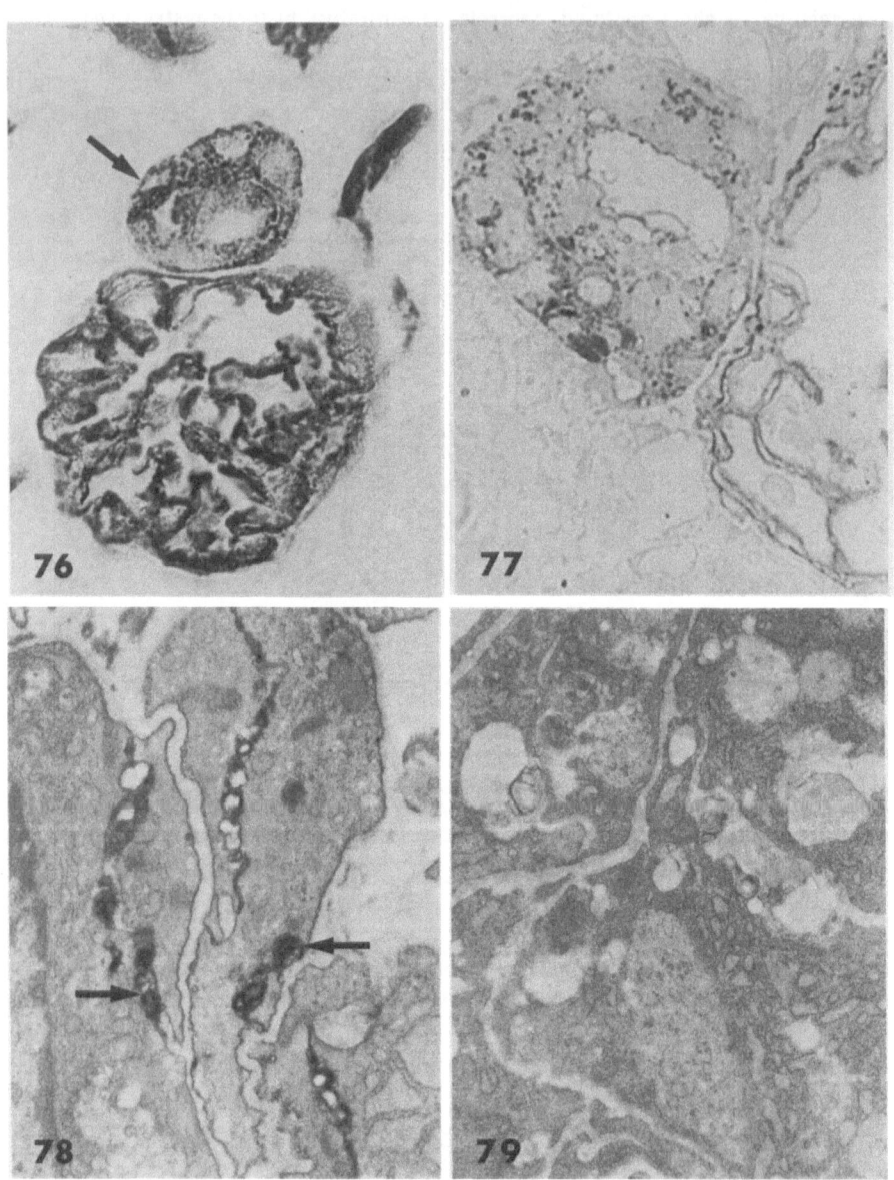

Both rats and mice show a decrease in brush border reactions (APA and APM) after adrenalectomy on qualitative histochemical study. In the rat this decrease is most apparent in the S_3 segment. However, in contrast to sham-operated animals, the vesicle-bound APA reactions show a marked increase in the S_2 segment and the beginning of S_3.

Biochemical Investigations

Fluorometric measurements of renal homogenate were performed only in rats. They indicate that the APA activities in adrenalectomized animals are significantly lower (U test, 99% level, Table 3) than in sham-operated controls — about 24% lower based on a comparison of mean values (Fig. 88). The APM activities in adrenalectomized animals also show a decline of about 13% compared with sham-operated controls (Fig. 88), but this is not statistically significant by the U test (95% level, Table 3).

4 Discussion

Our study focuses on two aminopeptidases occurring in the kidney, aminopeptidase A and aminopeptidase M, which were discovered by Pfleiderer and Celliers (1963, APM) and Glenner et al. (1962, APA). We know from biochemical studies that aminopeptidase A (APA; membrane aminopeptidase II, aspartate aminopeptidase, glutamate aminopeptidase, E.C. 3.4.11.7) is a membrane-bound enzyme (McDonald and Schwabe 1979, Kenny 1979 lit.) with a molecular weight of about 190 000 (Nagatsu et al. 1970). We do not yet know the subunits of which the enzyme may be composed. It is reported that APA specifically removes N-terminal α-L-glutamic acid residues and α-L-aspartic acid residues from peptides (McDonald and Schwabe 1979). Aminopeptidase M (APM; membrane aminopeptidase I, E.C.

Fig. 76. Adrenalectomized male mouse. Demonstration of APA. Incubation of 10-μm section in APA standard medium with HNF instead of FBB for 30 min at +30 °C. Besides the glomerulus, APA is demonstrated in a cross-section of the afferent arteriole, whose epithelioid cells are arranged in the form of a "polkissen" (*arrow*). × 500

Fig. 77. Adrenalectomized male mouse. Demonstration of APA in a semithin section. Specimen preparation as for ultracytochemistry. The enzyme was demonstrated by simultaneous azo coupling (α-L-Glu-MNA as substrate, HPR as coupling agent) at pH 6.5 by 120-min incubation at room temperature. The reaction pattern is comparable to that in Fig. 76 (polkissen). Reaction product is demonstrated at cell membranes (endothelial cells, podocytes) and in vesicular structures within epithelioid cells. × 1200

Fig. 78. Adrenalectomized male mouse. Ultracytochemical demonstration of APA in epithelioid cells. Specimen preparation as in Fig. 77. Reaction product is found at cell membranes (*dark lines*) and especially in areas of cell contact (*arrows*). × 14 000

Fig. 79. Adrenalectomized male mouse. Ultracytochemical control reaction for APA demonstration in epithelioid cells. Specimen preparation as in Fig. 77, but without substrate in the incubation medium. No reaction product is observed. × 14 000

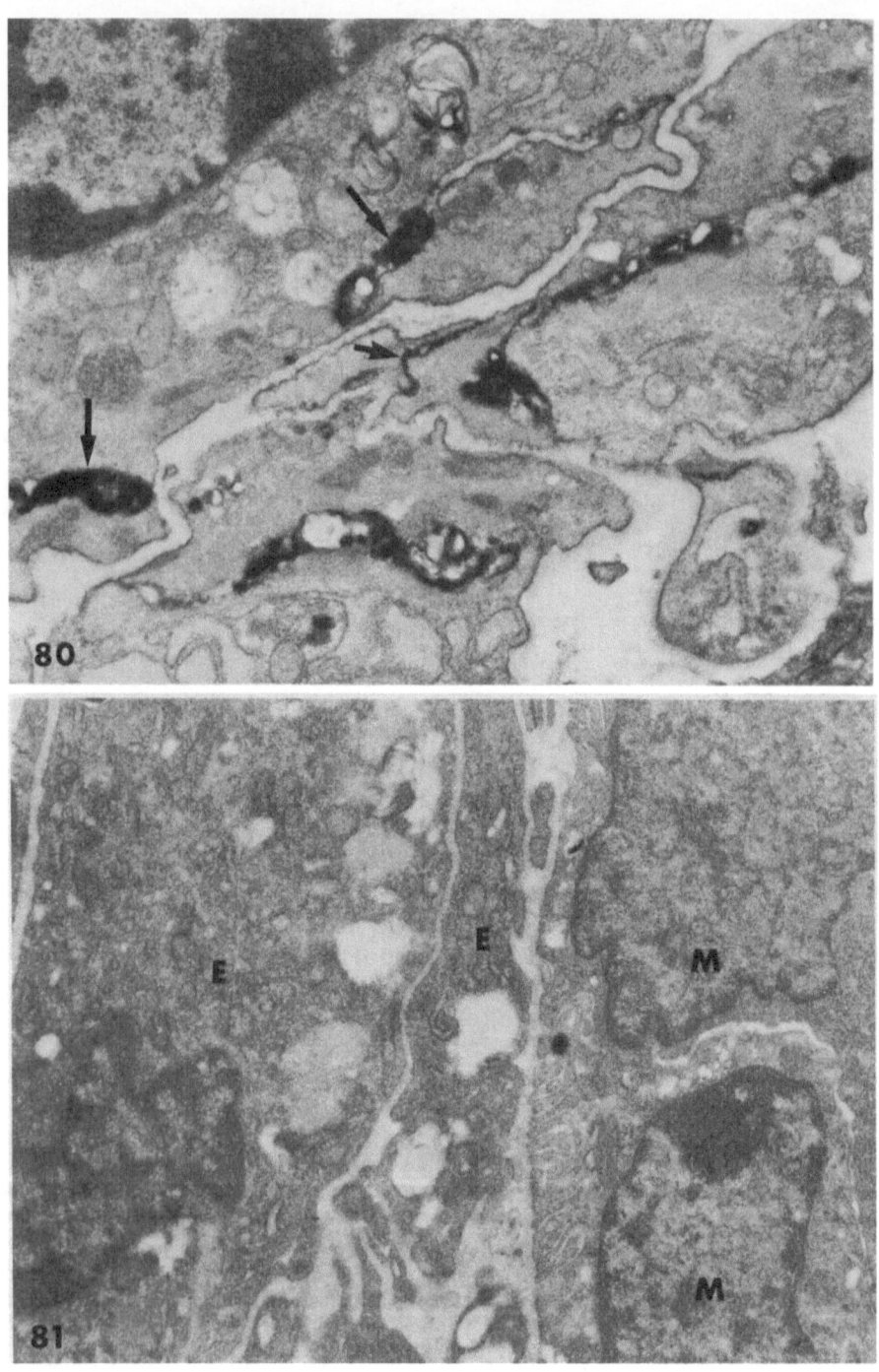

3.4.11.2) is also membrane-bound but, unlike APA, is a zinc-bearing glycoprotein that has a molecular weight of 240 000 (Scherberich et al. 1974) to 280 000 and reportedly is composed of five to ten subunits (Wachsmuth 1967, Lehky et al. 1973, McDonald and Schwabe 1979 lit.). APM is an aminopeptidase with a broad spectrum of activity which splits all N-terminal L-amino acid residues from peptides or poly-peptides, preferentially removes L-alanine, but also removes α-L-glutamic acid residues, α-L-aspartic acid residues, and L-proline residues at very low rates (Hanson et al. 1967, George and Kenny 1973, Wachsmuth and Donner 1976, Kenny 1979 lit; cf. Lojda et al. 1979).

The first step in our investigation was to optimize the procedure for the histo-chemical demonstration of APA. Because APA and APM also have a common distri-bution in the nephron, we differentiated the two enzymes on the basis of their diffe-rent ion dependencies. An important goal of our study, however, was to show histo-chemically and biochemically that both aminopeptidases take part in the degradation of angiotensins. We were able to confirm the identity of APA with a special angiotensi-nase known as angiotensinase A. One goal of the discussion is to clarify the function of the two aminopeptidases and their possible regulation based upon our localization studies and the results of animal experiments.

4.1 Methodological Aspects

Our methodological investigations indicate that simultaneous azo coupling with α-L-Glu-MNA as substrate and pure- or purest-grade FBB as coupler (cf. Lojda and Gossrau 1980) in acetone-fixed cryostat sections is best suited for the histochemical demonstration of APA activity. This results in the formation of an amorphous azo dye at the reaction sites, and the red reaction product provides a sharp color contrast for light microscopy. All of our histochemical investigations were performed with this method.

Below we shall discuss the essential aspects of this demonstration method. Briefly, they are: (a) pretreatment of the cryostat sections in acetone; (b) the use of substitut-ed amino acid-2-naphthylamides as substrates; and (c) the use of FBB as coupler. Then, based on our results on the calcium-ion dependence of APA, we shall discuss a means of differentiating between APA and APM, the identification of APA with angiotensinase A, and the relation of APM to angiotensin degradation in the kidney.

Pretreatment. First it should be noted that it is advantageous to demonstrate APA in cryostat sections of un-pretreated tissue. Block fixation in formaldehyde or glu-taraldehyde has proved unsuitable, for it leads to a marked activity loss in the ami-

Fig. 80. Adrenalectomized male mouse. Ultracytochemical demonstration of APA in epithelioid cells. The enzyme was demonstrated by simultaneous azo coupling (α-*L*-Glu-MNA as substrate, HNF as coupling agent) at pH 7 by 120-min incubation at room temperature. APA is demonstrat-ed at cell membranes (*dark lines*) and especially in areas of cell contact (*arrows*). ×21 000

Fig. 81. Adrenalectomized male mouse. Ultracytochemical control reaction for APA demonstra-tion in epithelioid cells. Specimen preparation as in Fig. 80, but without substrate in the incuba-tion medium. No reaction product was formed. *E*, epithelioid cells; *M*, cells of the macula densa. ×21 000

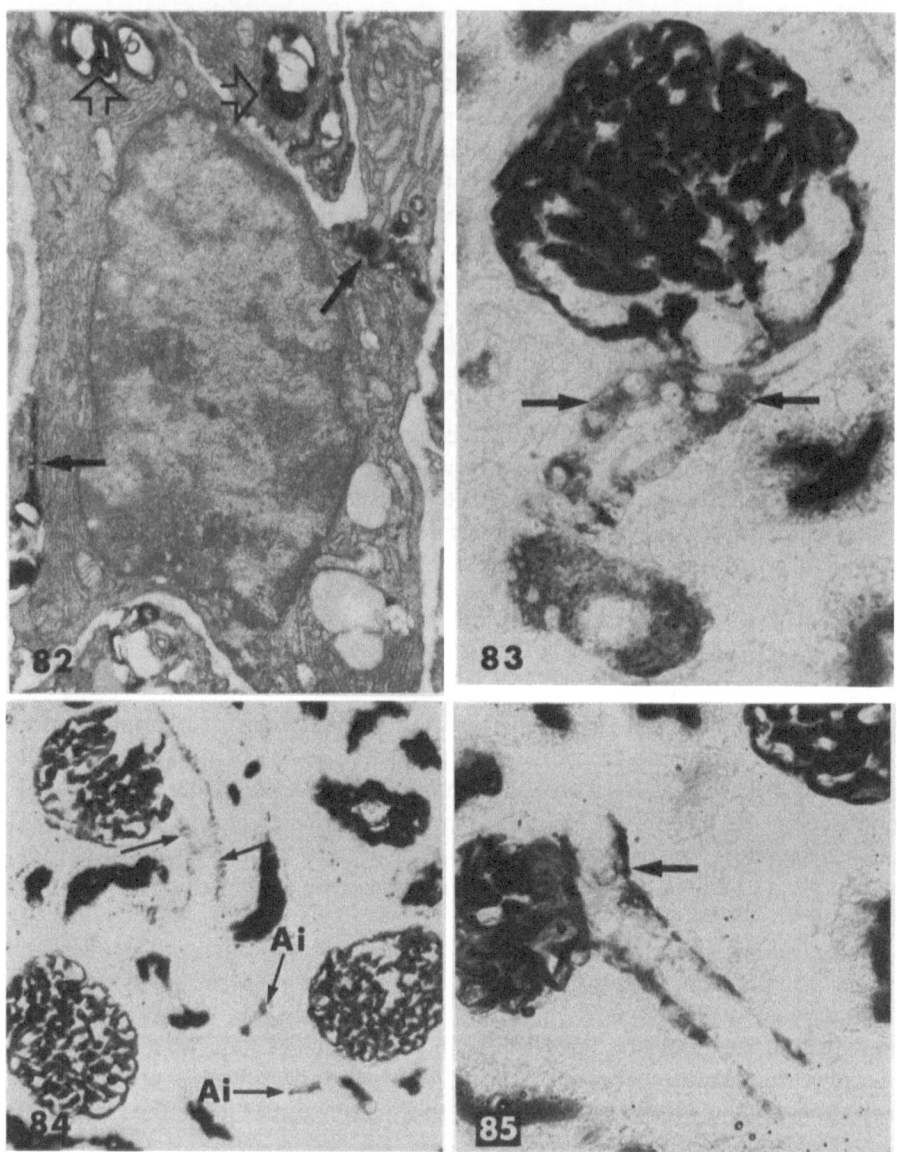

Fig. 82. Adrenalectomized male mouse. Ultracytochemical APA demonstration in epithelioid cells. The enzyme was demonstrated by simultaneous azo coupling (substrate α-L-Glu-MNA, coupling agent HPR) at pH 5 by 120-min incubation at room temperature. The reaction product is mainly localized in areas of cell contact (*solid arrows*) and in lysosomal structures (*open arrows*). ×10 500

Fig. 83. Adrenalectomized male rat. Demonstration of APA. Incubation of 10-μm acetone-pretreated section in APA standard medium for 20 min at +30 °C. Sectioned APA-positive epithelioid cells of the afferent arteriole (*arrows*) are seen in the vicinity of the glomerulus (with high APA activities). ×480

Fig. 84. Adrenalectomized male rat. Demonstration of APA. Specimen preparation as in Fig. 83. APA-positive epithelioid cells are found not only in the afferent arteriole (*arrows*) but also in the interlobular artery (*arrows, Ai*). ×150

Fig. 85. Adrenalectomized male rat. Demonstration of APA. Specimen preparation as in Fig. 83. The longitudinally sectioned afferent arteriole contains APA-positive epithelioid cells, mainly as individual elements. *Arrow*, Goormaghtigh's cell nest. ×350

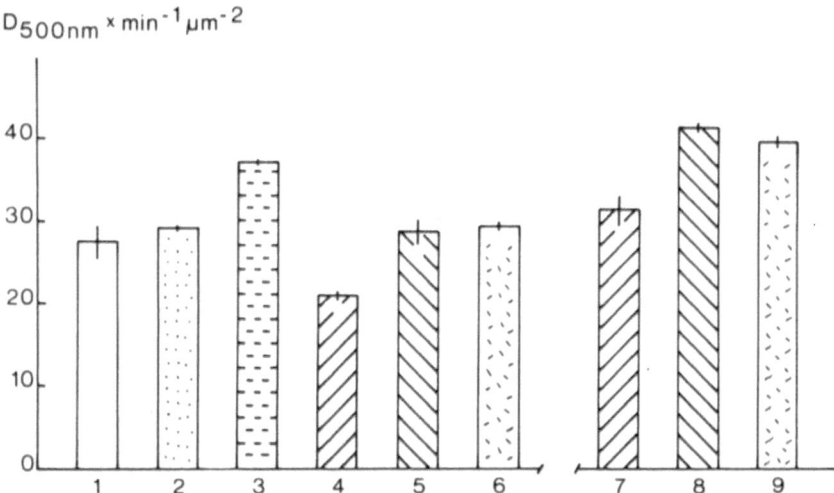

$D_{500nm} \times min^{-1} \mu m^{-2}$

Fig. 86. Quantitative histochemical determination of APA activities in subcapsular glomeruli of rats. The kinetic measurements of initial APA activities (*D*, relative density values) were performed with the APA quantifying medium (reaction temperature +30 °C, 5-μm renal sections). Each° column corresponds to a group of at least five animals, and at least five measurements were performed per animal. *Columns 1–6* = rats; *columns 7–9* = mice. *Column 1* = normal animals; *2* = sham operation; *3* = adrenalectomy; *4 + 7* = controls (Altromin diet C 1000); *5 + 8* = low-sodium diet (Altromin C 1036); *6 + 9* = high-sodium diet (Altromin C 1000 + 1% NaCl solution). Description of results in text

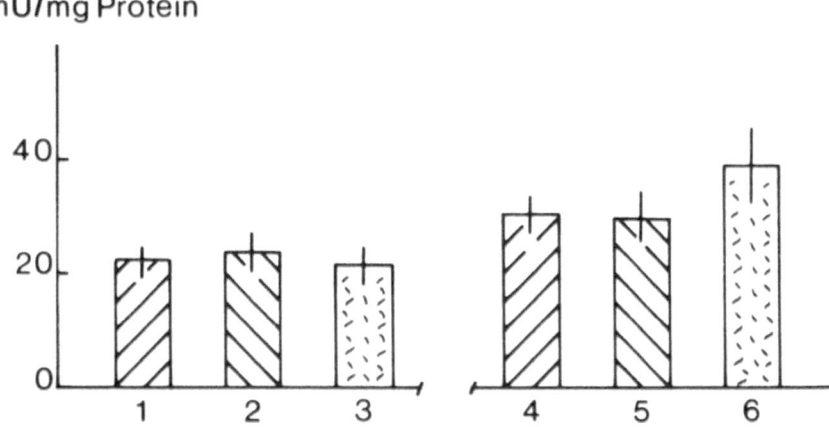

mU/mg Protein

Fig. 87. Determination of APA (*columns 1–3*) and APM activities (*columns 4–6*) in the renal homogenate of mice. The specific enzyme activities (mU/mg protein) were obtained by kinetic measurements of initial enzyme activities with the APA and APM fluorometric media in the ratio fluorometer (specifications in Fig. 32, reaction temperature +25 °C) and by determining the protein content (biuret method) of the renal homogenates in the spectrophotometer (Zeiss, PM 6). Each column corresponds to a group of at least five animals, and at least five measurements were performed per animal. *Columns 1 + 4* = control animals (Altromin diet C 1000); *2 + 5* = low-sodium diet (Altromin C 1036); *3 + 6* = high-sodium diet (Altromin C 1000 + 1% NaCl solution). Description of results in text

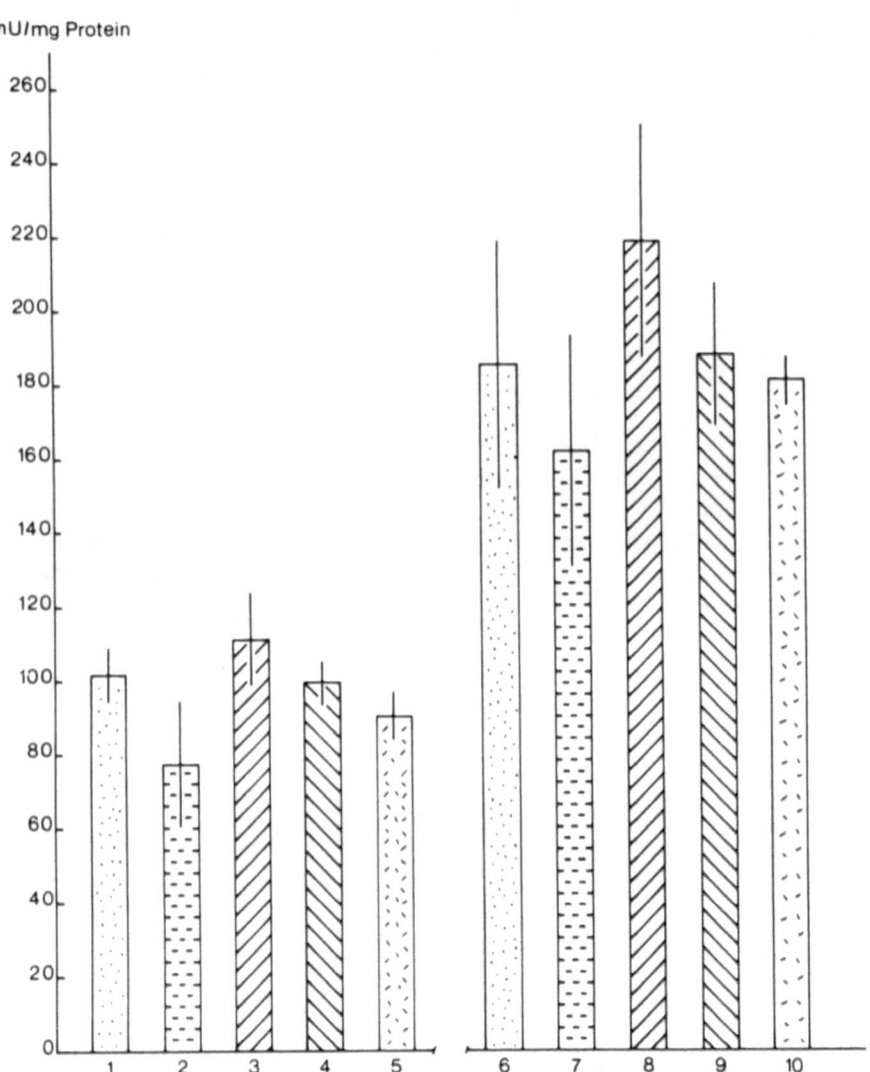

mU/mg Protein

Fig. 88. Determination of APA (*columns 1–5*) and APM activities (*columns 6–10*) in the renal homogenate of rats. The specific enzyme activities (mU/mg protein) were obtained by kinetic measurements of initial enzyme activities with the APA and APM fluorometric media in the ratio fluorometer (specifications in Fig. 32, reaction temperature +25 °C) and by determining the protein content (biuret method) of the renal homogenates in the spectrophotometer (Zeiss PM 6). Each column corresponds to a group of at least five animals, and at least five measurements were performed per animal. *Columns 1 + 6* = sham operation; *2 + 7* = adrenalectomy; *3 + 8* = controls (Altromin diet C 1000); *4 + 9* = low-sodium diet (Altromin C 1036); *5 + 10* = high-sodium diet (Altromin C 1000 + 1% NaCl solution). Description of results in text

nopeptidases under study. There is a particularly sharp decline of APA activities in glutaraldehyde-fixed material, as Lojda and Gossrau (1980) have shown.

Neither has the use of unfixed cryostat sections proved satisfactory, because with FBB as coupler, the azo dye is partially granular and is not stable enough for good localization. On the other hand, sharp localization of APA activities is obtained in freeze-dried cryostat sections that are mounted in a celloidin solution (FDC sections, Winckler 1970b). The demonstrable APA activity is very low with this method, however, so that reaction sites with weak activity are apt to be missed. APM is less strongly inhibited. This inhibition is probably due less to the freeze-drying than to the celloidin mounting (0.5% celloidin in ether, acetone and alcohol, Lojda et al. 1979; Kugler 1981b), since our own biochemical measurements have shown that APA activities in the homogenates of fresh kidneys and the homogenates of freeze-dried cryostat sections are approximately equal (unpublished results).

We achieved the best results with cryostat sections pretreated in acetone. Structural preservation is very good with this method, especially when the cryostat sections are fixed in 100% analytical-grade acetone at −25 °C without prior thawing. Acetone pretreatment is also important for degreasing of the sections, as the FBB coupler then yields a predominantly amorphous (non-granular) azo dye that is stable for several days, or even longer if the sections are stored at +4 °C after the reaction. The major drawback of acetone pretreatment is that it decreases the demonstrable APA activities, necessitating somewhat longer incubation times than in un-pretreated cryostat sections. Microdensitometric measurements in glomeruli have shown that acetone pretreatment reduces demonstrable APA activities by about 13% compared with untreated tissue sections. This may be due to an indeterminate enzyme inhibition or to an enzyme loss in the acetone; the cause remains unknown. In any case Lojda and Gossrau (1980) assume that the APA is not firmly membrane bound, as they report that varying amounts of enzyme go into solution during aqueous incubation. It is unclear, however, whether this escape of enzyme into the medium is due to weak structural binding of the APA or is a procedural artifact (e.g., membrane destruction by freezing and preparation of cryostat sections).

Our results with aldehyde-fixed cryostat sections (5 min in buffered formaldehyde or glutaraldehyde) are comparable in quality to those with acetone pretreatment. In contrast to block fixation (fixation time 20 h), there is only a slight decrease in demonstrable APA and APM activities in aldehyde-fixed cryostat sections. Localization with this method is only slightly inferior to that with acetone pretreatment.

Thus, on the whole we find that both the fixative and the fixation time are crucial factors in the preservation and localization of enzyme activities in the tissue section.

Substrates. The use of substituted 2-naphthylamide derivatives (α-L-Glu-MNA and α-L-Asp-MNA) is of considerable importance in the histochemical demonstration of APA in cryostat sections of un-pretreated tissue. Very good localization is obtained with these substrates. By comparison, the unsubstituted α-L-Glu-2NA provides an imprecise localization of APA acitivities, due to diffusion of the azo dye formed in the tissue section and to the precipitation of azo dye formed spontaneously in the medium.

A further difference between substituted and unsubstituted 2-naphthylamide derivatives is that α-L-Glu-2NA is hydrolyzed more rapidly by APA than is α-Glu-MNA. According to our photometric measurements in renal homogenate, the reaction rates of α-L-Glu-2NA are 40%–80% higher than those of α-L-Glu-MNA, depending on the buffer used (cf. Lojda and Gossrau 1980).

The substituted 2-naphthylamide derivatives α-L-Asp-MNA and α-L-Glu-MNA

61

also show differences in their hydrolysis rates. The enzyme activities demonstrated with α-L-Asp-MNA are generally lower, amounting to 60% of those with α-L-Glu-MNA, according to our microdensitometric measurements in glomeruli. Other investigators report that the hydrolysis rate of the unsubstituted α-L-Asp-2NA is 10–15 times lower than that of α-L-Glu-2NA (Glenner et al. 1962, Lojda and Gossrau 1980).

Coupling Agents. In addition to FBB, which served as our standard coupling agent, we also used Fast Garnet GBC (FGGBC), hexazonium-*p*-rosaniline (HPR), and hexazotized new fuchsin (HNF) for the demonstration of APA. We found that the unstable diazonium salts (HPR, HNF) inhibit the enzymic reaction more strongly then the stable diazonium salts (FBB, FGGBC) (cf. Lojda and Gossrau 1980). A further disadvantage of the unstable diazonium salts is that they form a yellow azo dye that contrasts poorly with the tissue (cf. Lojda and Gossrau 1980, Kugler 1981b), and reaction sites with weak activity may be overlooked.

Comparing the two stable diazonium salts FBB and FGGBC, we find that FBB is the preferred coupling agent, for the azo dye formed by the reaction is more or less amorphous in light microscopy. On the other hand, FGGBC, a commonly used coupling agent in photometric studies (Glenner and Folk 1961, Glenner et al. 1962, Nagatsu et al. 1970), yields a product that is more granular under the light microscope. Our qualitative histochemical studies indicate that for an optimal demonstration of APA, the best concentration of coupling agent in the reaction medium is 1 mg/ml for pure-grade FBB, 0.5 mg/ml for purest-grade FBB, and 2–2.5 mg/ml for pure-grade FGGBC. Thus, when FGGBC is used for azo coupling, relatively high concentrations are needed. Other investigators recommend considerably lower concentrations of FGGBC (0.37 mg/ml: Glenner and Folk 1961; Glenner et al. 1962) and higher concentrations of purest-grade FBB (1 mg/ml: Lojda and Gossrau 1980) than our results would indicate.

On the basis of these results, we are convinced that in our case FBB is the coupling agent of choice. For the assessment of our quantitative histochemical results, we considered it important to determine more accurately the degree of APA inhibition caused by FBB, employing biochemical photometry for this purpose. We found a mean inhibition rate of between 2% (0.25 mg/ml purest-grade FBB) and 11% (1 mg/ml purest-grade FBB) relative to control measurements (without purest-grade FBB in the incubation medium during the reaction). Lojda and Gossrau (1980), on the other hand, report that 1 mg/ml purest-grade FBB inhibits APA activity by 75%.

This large discrepancy is apparently based upon procedural differences leading to differences in spontaneous azo dye formation or in the stability of the azo dye that is formed. We say this because, as we have found, photometric azo dye measurements require highly accurate timing in the individual incubation steps. This is necessary because spontaneous azo dye formation is relatively great during the reaction, and because the azo dye is not stable after the enzyme reaction is stopped, i.e., the extinctions undergo a nearly linear decline after a very short time. This decline is relatively slight in the aqueous media used by us but is large in ethyl acetate (unpublished results), which is often used for elution of the azo dye (cf. Lojda and Gossrau 1980).

Ion Effects. The biochemical differentiation of APA and APM is an extremely difficult matter. According to previous studies, the two enzymes have practically the same localizations in the cell membrane, and attempts to differentiate them with biochemical methods have had only limited success (Nagatsu et al. 1970, Kenny 1979). This presents difficulties for the immunocytochemical visualization of APA, because a specific antibody against APA can be obtained only by isolating the enzyme in pure form. However, the dependence of APA on calcium ions has been suggested as a possible mechanism for differentiating this enzyme (Glenner and Folk 1961, Glenner et al. 1962, Hess 1965, Nagatsu et al. 1970, McDonald and Schwabe 1979, Kenny 1979, Lojda and Gossrau 1980, Kugler 1981b).

This prompted us to investigate the ion dependence of APA. Using biochemical, qualitative histochemical, and quantitative histochemical methods, we were able to confirm beyond question the calcium-ion dependence of APA activities. Qualitative histochemistry shows that APA activity is promoted by calcium ions at all reaction sites in the kidney, the acitivities increasing with the calcium ion concentration (1.5, 5, 10 mM). Our quantitative histochemical work on the glomerulus also confirms this. The highest activities are measured at a concentration of 10 mM CaCl$_2$, which leads to a 2.5-fold increase in APA activities.

Biochemical fluorometric APA measurements in renal homogenate as a function of calcium concentration yield essentially the same results. Here, however, an activity increase is observed only up to 5 mM CaCl$_2$ in the incubation medium; this increase is shown better with α-L-Asp-2NA as the substrate than with α-L-Glu-2NA. With α-L-Asp-2NA, a concentration of only 1.5 mM CaCl$_2$ produces a three fold increase in APA activities. With 10 mM CaCl$_2$ APA activities begin to decline again according to fluorometric measurements. These findings largely agree with Glenner et al. (1962, cf. Kenny 1979), who found that APA activities increased 2.4-fold with 10 mM CaCl$_2$ and then fell somewhat with 20 mM CaCl$_2$.

Thus, our biochemical and histochemical studies clearly indicate that APA is acitivated by CaCl$_2$, and that, in agreement with Glenner et al. (1962, cf. Kenny 1978), this activation is maximal in the range of 5–10 mM CaCl$_2$, where a two- to threefold increase in APA activities is observed.

EDTA, which forms complexes of divalent cations (e.g., Ca^{2+}), is an inhibitor of APA. We were able to demonstrate this inhibition clearly in our qualitative histochemical and biochemical preparations, but less clearly in our quantitative histochemical studies of glomeruli. With 1.5 mM EDTA in the renal homogenate (α-L-Asp-2NA as substrate), only 35% of the original APA activities (without CaCl$_2$) can be demonstrated on biochemical fluorometric analysis. This inhibition increases as the EDTA concentration is increased to 10 mM, but only to a slight degree; thus, we obtained no complete inhibition of APA like that described by Glenner and Folk (1961) with 10 mM EDTA. This may derive from the fact that our fluorometric measurements are more sensitive than the photometric methods with FGGBC used by Glenner and Folk (1961).

Experiments on the activation and inhibition of APA with calcium and EDTA, respectively, indicate that the effects of the divalent cation and complexing agent are strongly dependent on the buffer used (unpublished results). Thus, our findings cannot be directly compared with the data of other authors, or our own data – where various buffers were used – are not directly comparable with one another.

While APA demonstrates a clear dependence on calcium ions, this is not the case with APM. We tested the ion dependence of APM with qualitative histochemical and biochemical fluorometric methods. Qualitative histochemistry revealed no effects from calcium ions (1.5, 5, and 10 mM) or EDTA (1.5, 5, and 10 mM) relative to the controls. Biochemical analysis showed that APM activity is largely unchanged up to 5 mM CaCl$_2$, which agrees with the findings of Nagatsu et al. (1970) in relatively pure APM preparations. With 10 mM CaCl$_2$, however, APM activity is increased by about 15%. This may result from calcium-substrate interactions at this high CaCl$_2$ concentration which promote access of the artificial substrate to the active center of the APM. Biochemical analysis showed no significant APM activity changes with 1.5, 5, and 10 mM EDTA. Our results on EDTA agree with published findings that neither 2 mM (Lojda and Gossrau 1980) nor 50 mM EDTA (Hess 1965) inhibits APM. It has, however, been reported that APM is inhibited by preincubation in EDTA (Pfleiderer et al. 1964).

We know, then, that APA and APM show distinct differences in their activation by calcium ions and inhibition by EDTA, and thus that calcium ions provide a means of differentiating the two enzymes.

In connection with our ion-dependence studies, we also investigated the influence of various NaCl concentrations (50–180 mM) and buffers (tris-maleate, sodium cacodylate, tris-HCl) on APA activities using different methods (partly quantitative histochemical, partly biochemical). According to our quantitative histochemical measurements in glomeruli, 100 and 130 mM NaCl in the incubation medium increase APA activities by approximately 17%. With NaCl concentrations below 100 mM or above 130 mM, however, APA activities start to decline, 180 mM NaCl causing a mean inhibition of 5% compared with controls (without NaCl). These findings are difficult to interpret, however, because possible interactions of the buffer with the NaCl cannot be excluded. On the other hand, the results could explain why the APA acitivities determined by quantitative histochemistry in glomeruli with 0.1 M sodium cacodylate (pH 7) are approximately 14% higher than with tris-maleate buffer. The activating effect of the cacodylate buffer could derive from its high sodium content.

We also investigated the effect of various buffers on APA activities by fluorometric and photometric measurements in renal homogenate. The photometric technique involved simultaneous azo coupling of the 4-methoxy-2-naphthylamine and 2-naphthylamine liberated from α-L-Glu-MNA and α-L-Glu-2NA, respectively, by APA activity with purest-grade FBB. Formation of the color product was measured kinetically at 525 nm. Significantly higher APA activities can be measured in the cacodylate buffer than in tris-maleate buffer. We determined the APA activities fluorometrically with 0.1 M tris-HCl and 0.1 M cacodylate buffer, having first constructed standard curves of 2-naphthylamine fluorescence in the corresponding buffers. Both the standard curves and APA activities yield identical results with both buffers. This conflicts with the findings of Lojda and Gossrau (1980), who observed the highest APA hydrolysis rates with cacodylate buffer.

Identity of APA and Angiotensinase A. According to biochemical studies by Hess (1965) and Nagatsu et al. (1970), APA is identical to angiotensinase A, and is thus an enzyme which cleaves N-terminal aspartic acid from ANG I and II. Until now, however, this identity has not been proven histochemically.

To clarify the relation of APA to angiotensinase A, we determined the effect of various angiotensins on glomerular APA using quantitative histochemical methods (APA inhibition tests). Our quantitative technique involved enzyme-kinetic measurements, i.e., continuous measurements of the reaction product formed by APA activity at +30 °C. A specially developed measuring unit was utilized for this purpose (Kugler 1981a).

Before testing the inhibition of APA by the angiotensins, we first tested the usefulness of the method. In these preliminary studies we determined the absorption maximum of the red azo dye formed by APA in the renal section with α-L-Glu-MNA as substrate and purest-grade FBB as coupling agent. We found that absorption is maximal at 500 nm (cf. Lojda and Gossrau 1980). We also measured APA as a function of section thickness (i.e., at various APA concentrations) and of reaction time, using the glomerulus as the measuring site. With regard to APA measurements as a function of section thickness, a linear relationship is observed between the formation of reaction product and section thickness (3–12 μm), thus satisfying a key prerequisite for quantitative histochemical measurements (Wachsmuth 1980, Kugler 1981a). APA measurements as a function of reaction time show that the reaction is linear for only about the first 1–2 min. This was revealed by the good time resolution of our special instrument setup, which performed measurements at 6-s intervals. Our results conflict with the findings of Lojda and Gossrau (1980), who measured a 15-min period of reaction linearity. This descrepancy is very likely due to differences of method. The brief period of linearity is probably the result of a "clogging effect" as the azo dye blocks the diffusion of substrate to the active center of the APA. Similar findings were made in quantitative histochemical measurements of succinate dehydrogenase with a tetrazonium salt procedure (Kugler 1981a). It thus appears that endpoint measurements beyond a reaction time of 1–2 min may lead to false and incomparable results. We always determined initial enzyme activities in our measurements, therefore.

Microdensitometric measurements to determine APA activities under the influence of ANG I, II, and III were always performed on subcapsular glomeruli. The glomerulus was chosen as the measuring site because it contains no APM (cf. Sect. 3.2; Wachsmuth 1968, Wachsmuth and Torhorst 1974, Wachsmuth and Donner 1976, Kugler 1981b) or other peptidases that are known to interact with APA in angiotensin degradation. This also applies to the dipeptidyl peptidase IV (DAP IV) occuring in the glomerulus, which, according to our findings, is not inhibited by angiotensins (unpublished results). We used angiotensins in which isoleucine occurs at position 5 of the angiotensin molecule. This type of angiotensin occurs physiologically in the rat as well as in most other species investigated to date (Sokabe 1974, Nishimura 1980).

First we determined the K_m of APA in the rat glomerulus, obtaining a value of 0.23 mM. All the angiotensins (I, II, and III) competitively inhibit APA in rat and mouse glomeruli (plot after Lineweaver and Burk 1934); the K_i of ANG II in the rat glomeruli is 0.14 mM.

Thus, it has been confirmed by quantitative histochemical analysis that histochemically demonstrable APA is identical to angiotensinase A. The competitive inhibition, which is characterized in the Lineweaver-Burk plot by a common intersection of the substrate-dependent lines and the angiotensin-produced inhibition lines on the l/V axis, means that the angiotensins, as natural substrates of APA, compete with the artificial substrate α-L-Glu-MNA for the APA active center.

But the substrate affinity is apparently not based only upon the N-terminal aspartic acid of ANG I and II, because ANG III, which lacks N-terminal aspartic acid, also competitively inhibits APA. Other investigators (Hess 1965) report that even when the angiotensin molecule is shortened further (Val-Tyr-Val-His-Pro-Phe), it still

65

competes for the active center of APA (microsomal fraction of the rat kidney). However, as our qualitative histochemical and biochemical fluorometric studies (see below) show, the angiotensin residue His-Pro-Phe also appears to compete with artificial substances for the APA active center, but not the angiotensin fragment Tyr-Ile.

To check our quantitative histochemical findings in rat glomeruli, we determined APA activities in microdissected glomeruli by means of biochemical fluorometric analysis. These fluorometric measurements in glomerular homogenates indicate a K_m of 0.23 mM and for ANG I and II a competitive inhibition. Because these measurements were done with an unsubstituted 2-naphthylamide (α-L-Glu-2NA) and with cacodylate buffer, the histochemically (azo coupling) and fluorometrically determined K_m values are not entirely comparable, although the same value was obtained with both methods. The data obtained in microdissected glomeruli confirm the adequacy of the histochemical quantifying method, however.

For control purposes we used both methods — fluorometry in microdissected glomeruli and quantitative histochemistry in glomeruli of renal sections — to measure the reaction of L-Ala-2NA and L-Ala-MNA respectively, in order to demonstrate any APM activities that might by present. Quantitative histochemistry revealed an extremely low reaction rate for L-Ala-MNA, indicating an APA-to-APM ratio of approximately 69:1. Thus, APA occurs in the glomerulus in relatively pure form. The APM activities still measurable in the glomeruli may have derived from blood constituents in the glomerular capillaries, since the kidneys were not perfused. This appears likely because blood contains APM and APA (Khairallah et al. 1963, Hess 1965, Doyle et al. 1967, Khairallah and Page 1967, Kurtz and Wachsmuth 1969, Nagatsu et al. 1970, Ledingham and Leary 1974 lit., Ryan 1974 lit., Khairallah and Hall 1977 lit.). Thus, "pure" APA for immunocytochemical purposes could presumably be obtained from rat glomeruli.

The situation was different with the microdissected glomeruli, in which we were able to demonstrate relatively high APM activities, although the kidneys had been perfused and thus largely purged of blood constituents. The ratio of APA to APM was 6:1. This may be attributed to our procedure, in which the renal sections were freeze-dried in "stacks," allowing glomeruli to become contaminated with brush border fragments. Thus, the quantitative histochemical investigations represent the relatively "purest" conditions for APA measurements.

Having established that APA is angiotensinase A, we performed qualitative and quantitative histochemical studies to determine the effect of angiotensins on APA at all reaction sites in the nephron. In the qualitative studies we observed a marked inhibition of APA by ANG I, II, and III not only in the glomerulus, but also in the other well-known reaction sites of the kidney, namely in the brush borders of the proxima tubule (Glenner et al. 1962, Lojda and Gossrau 1980, Kugler 1981b). Comparing ANG II and III, we find that ANG II is the stronger inhibitor of APA. According to our quantitative histochemical measurements in the nephron of the male rat with substrate-equimolar ANG II concentrations (1.5 mM), there is a 60% inhibition of APA activities both in the glomerulus and in the brush borders of the S_3 segment Besides the high inhibition rate, the equally high inhibition of APA in the glomerulus and S_3 segment is noteworthy.

Relation of APM to Angiotensin Degradation. APM is presumed to take part in angiotensin degradation (Hess 1965, Nagatsu et al. 1970, Lojda and Gossrau 1980), because it is a relatively substrate-nonspecific remover of amino acids from peptides. Until now, however, the relation of APM to angiotensin degradation in the kidney has not been subjected to accurate biochemical or any histochemical studies. To explore the involvement of APM in angiotensin degradation, we performed qualitative histochemical investigations and biochemical measurements in renal homogenate with ANG I, II, and III.

Our qualitative studies with ANG I, II, and III showed a marked inhibition of APM at the aforementioned reaction sites in the kidney (brush borders of the proximal tubule), a comparison of ANG II and III indicating a stronger inhibition by ANG III.

With regard to the biochemical fluorometric measurements in whole kidney homogenate, it must be considered that we were working with mixed-rather than pure-enzyme preparations. Such homogenate contains brush-border peptidases as well as lysosomal peptidases (Kenny 1979, Jedrzejewski and Kugler, to be published) that are capable of degrading angiotensins (Regoli et al. 1963, Khairallah and Page 1967, Yang et al. 1968, Johnson and Ryan 1968, Matsunaga 1971), though available data on the rat kidney are fragmentary. The brush borders in particular contain a number of peptidases in close proximity to one another that develop, or might well develop, hydrolytic activities toward angiotensin molecules, such as endopeptidases (Kenny 1979), APA (Glenner et al. 1962, Hess 1965, Nagatsu et al. 1970, George and Kenny 1973, Kenny 1979, Lojda and Gossrau 1980, Kugler 1981b), APM (George and Kenny 1973, Kenny 1979, Lojda et al. 1979, Kugler 1981b), and converting enzyme (identical to kininase II; Ward et al. 1975, 1976; Ward and Erdös 1977, Taugner et al. to be published). This makes it difficult to interpret the results of inhibition experiments with angiotensins. For purposes of comparison we performed such experiments not just with the APM contained in the homogenate, but also with APA.

Using the Lineweaver-Burk plot, we determined a K_m of 0.13 mM for APA and 0.24 mM for APM. These values are nearly identical to those obtained by Hess (1965) in microsomal fractions of the rat kidney. We conducted our inhibition experiments with ANG I, II, and III using various substrate concentrations with two fixed angiotensin concentrations. We found that ANG II competitively inhibits APA (with K_i = 0.015 mM), while ANG III competitively inhibits APM (K_i = 0.003 mM). This K_i of APA on ANG II inhibition is somewhat lower, though still of the same order of magnitude, as that reported for APA by Hess (1965) in microsomal fractions of the kidney (0.045 mM) and for angiotensinase A by Klaus et al. (1963) in human serum (0.06 mM). However, as in our homogenate studies, microsomal fractions of the kidney do not represent pure-enzyme preparations.

ANG I and III noncompetitively inhibit APA, while ANG I and II noncompetitively inhibit APM. Taking into account our quantitative histochemical results on the inhibition of glomerular APA by angiotensins, two possible conclusions may be drawn: (a) Converting enzyme, APA, and APM are localized in such close proximity to one another that they influence the active centers of the neighbouring enzymes through intermolecular changes in the form of multiple-substrate reactions (Bisswanger 1979), so that the inhibition of APA and APM is competitive or noncompetitive depending on the angiotensin used. (b) Various peptidases degrade the angiotensins simultaneously, giving rise to peptide fragments that create a confusing inhibitory picture. The latter hypothesis appears more likely, as long as the foregoing enzymes cannot be demonstrated in pure form.

Our fluorometric measurements of renal homogenate, in which the rates of APA and APM inhibition by L-Tyr-Ile and L-His-Pro-Phe were determined, are also of interest in this context. While both peptides, which are fragments of the ANG II molecule, inhibit APM by about 25% in substrate-equimolar amounts, only the tripeptide inhibits APA to a comparable degree. This suggests that certain portions of the angiotensin molecule are important for binding to the active center of APA (see above).

On the whole, our inhibition experiments with angiotensins allow us to conclude that APM and APA are both involved in angiotensin degradation.

4.2 Animal Experiments

APA and APM are histochemically demonstrable mainly at three sites in the rat and mouse nephron: the juxtaglomerular apparatus (JGA), the renal corpuscle, and the proximal tubule. The localizations of the enzymes show basic differences. While APA is present in all three parts of the nephron, APM occurs only in the proximal tubule (cf. Kugler 1981b).

Below we shall consider APA and APM only as they relate to angiotensin degradation. Of course APM in particular hydrolyzes many other peptides owing to its broad spectrum of activity. Because both aminopeptidases possess somewhat different localizations and activities in the nephron, angiotensin degradation varies from one part of the nephron to the next, as does the local supply of intact angiotensin. Physiologic studies indicate that angiotensins perform different tasks in the JGA, glomerulus, and proximal tubule (Hierholzer 1977). The angiotensin receptors in these structures may well play a role in this process (Sraer et al. 1974, Beaufils et al. 1976, Freedlender and Goodfriend 1977). A finding in the rabbit aorta seems to have an important bearing on these receptors (Kalsner and Nickerson 1968, cf. Peach and Levens 1980): It was shown that angiotensin-degrading enzymes are located directly adjacent to angiotensin receptors. This finding lends special importance to our studies on the distribution of angiotensin-degrading enzymes in the kidney, since the localization of these enzymes may mark the site of these receptors.

4.2.1 Juxtaglomerular Apparatus

Microdissection studies have demonstrated the presence of angiotensinases in the JGA (Granger et al. 1969, Thurau et al. 1970, Granger et al. 1972, Dahlheim and Schmid-meier 1975). Until now it has not been possible to specify these angiotensinases or accurately describe their localization. Using histochemical methods, we have succeeded in demonstrating a special angiotensinase, APA (angiotensinase A), in the JGA of the rat and mouse. As we were able to show with histochemical and biochemical methods, this enzyme splits off aspartic acid, i.e., the N-terminal amino acid of ANG I and II. In addition, we have determined the precise distribution pattern of this peptidase in the JGA of the rat and mouse. It is different in the two species. In rats the Goormaghtigh's cells, and sometimes vascular segments near the glomerulus, show a positive reaction. Adrenalectomized rats also show APA activities in the epithelioid cells of the afferent arteriole, however. This suggests that APA also occurs there under normal circumstances but is below the limit of histochemical detection. In the mouse, by contrast, APA is demonstrable in the epithelioid cells of normal animals, where it shows varying degrees of activity. The Goormaghtigh's cells of the mouse appear to be negative on the whole, although the poor development of the Goormaghtigh's cells nests in the mouse (cf. Gorgas 1978) makes them difficult to distinguish from surrounding structures. The macula densa is APA-negative in both species.

For a functional interpretation of our results, we proceed from the assumption that APA hydrolyzes angiotensins that are formed locally in the JGA or are carried there by the blood or lymph, and thus alters the local angiotensin supply. The synthesis of ANG II in the JGA itself (Schmid 1962, Thurau 1964, McGiff 1968, Britton 1968, Thurau 1973, 1974, 1975 lit; Stowe et al. 1979) is indicated by the fact that renin substrate (α_2-globulin; Sutherland 1970, Finkielman et al. 1972) and converting enzyme (Granger et al. 1969, Thurau et al. 1970, Granger et al. 1972, Leckie et al. 1972) have been demonstrated or probably occur there. Local ANG II production is further evidenced by the fact that renin is secreted not only into the bloodstream, but also into the renal lymph (Lever and Peart 1962, cf. Horky et al. 1971, Thurau 1975), and converting enzyme is also present there (Lever and Peart 1962, cf. Horky et al. 1971). This further suggests that the intercellular space of the JGA and especially of the Goormaghtigh's cell nests also contains renin and converting enzyme, thus providing substrates and enzymes for local ANG II formation in the JGA.

The cellular compartments of the JGA in which angiotensin occurs or is presumed to occur and the functions that ANG II performs in the JGA remain to be discussed. Perhaps the localization and significance of APA can then be correlated with the distribution and function of angiotensin.

Little is known as yet about the cellular compartmentalization of ANG II in the JGA itself. However, receptor studies in various organs have shown that ANG II is mainly bound to cytoplasmic membranes (Peach and Levens 1980). The occurence of substrates and enzymes of the renin-angiotensin system in the renal lymph (see above) also suggests that angiotensin first reaches the cytoplasmic membranes of cells of the JGA. This consideration has a bearing on our ultracytochemical findings on APA, for this enzyme occurs chiefly in cytoplasmic membranes of the described JGA structures. If we assume that angiotensin exerts an effect on cell membranes, then the APA demonstrated there could regulate this effect through the breakdown of ANG II. However, our ultracytochemical studies have shown that APA also occurs intracellularly in myelin-like laminar, vacuolar, and lysosomal structures in the epithelioid cells of mice. Such structures are thought to be associated with the maturation or dissolution of renin granules (Lee et al. 1966, Bucher et al. 1967, Rouiller and Orci 1971, Bucher and Kaissling 1973 lit.). Because ANG II occurs in the cell organelles of various organs (Robertson and Khairallah 1971, Peach and Levens 1980 lit.), it would appear that APA performs an intracellular function in angiotensin degradation. Such a function of APA in the JGA is also indicated by the fact that ANG II could be demonstrated immunocytochemically in epithelioid cells of the rat kidney (Taugner et al. to be published, Celio and Ingami 1981), but not in the mouse kidney (Taugner et al. to be published). Perhaps our APA findings in rat and mouse epithelioid cells explain this species difference — for the APA demonstrable in the epithelioid cells even of normal mice may hydrolyze angiotensin so rapidly that insufficient angiotensin is available for demonstration. The opposite situation is observed in the rat, in which the APA activity of epithelioid cells in the normal animal is below the limit of histochemical detection. As a result, angiotensin is degraded slowly in these cells, and ANG II can be demonstrated immunocytochemically.

While our knowledge of the cellular compartmentalization of ANG II in the JGA is so far limited to the fact that it occurs intracellularly, more is known about the action of ANG II in the JGA. ANG II exerts a duel function there, on the one hand inhibiting renin secretion (Hierholzer 1977), and on the other stimulating the contraction of myoid elements of the JGA (ANG II is a potent agonist of smooth muscle contraction, cf. Bohr 1974 lit.).

We shall next consider the relationship between ANG II, renin production, and the occurrence of APA. While ANG II inhibits renin secretion, APA counteracts this hormonal signal by its degradation of ANG II, suggesting that the local supply of ANG II and APA activity in the JGA determine the level of active ANG II.

Our findings in normal and experimental animals are consistent with this assumption. Thus, we found in rats that APA-positive Goormaghtigh's cell nests and APA-positive vascular segments near the glomerulus are more numerous in subcapsular glomeruli than in juxtamedullary glomeruli. The functional distribution of renin-producing granular epithelioid cells shows a similar pattern. According to Rouffignac et al. (1974), more renin is produced in the subcapsular JGA than in the juxtamedullary JGA; and according to Faarup (1971), the granularity of superficial JGAs is higher than that of JGAs lying deeper in the cortex. Assuming that ANG II inhibits renin secretion, the increased APA activities of subcapsular JGAs would reduce the local ANG II concentration, resulting in a greater production of renin in these areas. On the other hand, it is also conceivable that the increased production of renin, and thus of ANG II, in subcapsular JGAs is responsible for the higher APA activities demonstrated at these sites.

Our findings in experimental animals are important in clarifying the relationship between APA activity and renin production, particularly those involving adrenalectomized animals. Although we have not personally investigated renin in the blood and kidneys of adrenalectomized animals, it has been proved by other researchers that increased amounts of renin are secreted after adrenalectomy (Gross and Sulzer 1957, Schaechtelin et al. 1963, Gross et al. 1965, Danda and Deveny 1971, Bucher and Kaissling 1973 lit.). This is also indicated by immunocytochemical studies of semithin sections of rat and mouse kidney which show that more renin is demonstrated in epithelioid cells of the JGA after adrenalectomy than in control animals (Davidoff and Schiebler 1981). On the other hand, this increase in renin production leads to an increased synthesis of ANG II. Our findings in rats and mice indicate that adrenalectomy leads to increased APA activities in epithelioid cells of both the afferent and efferent arteriole as well as in the interlobular artery. The increased hydrolysis of ANG II associated with the high APA activities means that the APA counteracts the inhibition of renin secretion by high ANG II concentrations in the JGA. As a result, renin secretion is increased after adrenalectomy.

Besides its effect on renin secretion, ANG II may perform an additional function in the JGA — that of regulating the glomerular blood flow (Thurau 1973, 1974, 1975). Thus, it is reported that ANG II leads to a contraction of JGA structure. Contractile properties apparently reside in epithelioid cells and smooth muscle of the afferent and efferent arterioles as well as in Goormaghtigh's cells, which contain myofilaments (Gorgas 1978 lit.). The vascular lumen of the afferent and efferent arteriole, and thus the glomerular blood flow, could be influenced by tonic contractions of these structures. Moreover, an ANG II-mediated calcium influx into the cells of these structures could have a special impact on the contractile state of the JGA. Studies on muscle preparations have demonstrated that the triggering of a contraction by angiotensin is dependent upon calcium influx into the cell. It is assumed that ANG II mediates this calcium flow (Peach and Levens 1980). Membrane-bound APA might well modify this angiotensin-controlled calcium flow, and thus the contractile state, through the degradation of ANG II. It is significant in this regard that the various extramacular cells of the JGA have intensive contacts in the form of nexus and gap junctional connections (Gorgas 1978), prompting Gorgas (1978) to regard them as a functional unit. This presumes an electronic coupling of these cells, so that synchronous contractions can take place in the JGA region (Gorgas 1978). Our ultracytochemical finding of increased APA activities in the region of cell contacts is of interest in this regard. Because impulse conduction is presumably enhanced in this region (Gorgas 1978), intercellular calcium shifts could be mediated by ANG II and regulated by varying APA activities.

Owing to their central location between the macula densa and vascular pole of the glomerulus, Goormaghtigh's cells are presumed to exercise a special function in the JGA (cf. Kriz and Taugner 1977). They could have special significance in the transmission of ionic signals from the macula densa to the vascular pole or in the control of glomerular blood flow. Thus, besides the APA that occurs in the Goormaghtigh's cells of the rat, other histochemically demonstrable enzymes are of importance in these cells (Kugler, unpublished work). Specifically, we have been able to demonstrate alkaline phosphatase (cf. Kroon 1960, tetrazolium method of McGadey 1970) as well as magnesium-dependent ATPase (lead salt method of Von Deimling 1964) and calcium-dependent ATPase (calcium-cobalt method of Padykula and Herman 1955) in the Goormaghtigh's cells. These enzymes also show reactions of varying degree in the region of the afferent and efferent arteriole. These hydrolytic enzymes point to the occurrence of contractile processes and ion-transport processes in this region.

Summarizing our results and considerations on APA in the JGA, it would appear that this enzyme modifies angiotensin effects in the JGA through angiotensin degradation. The APA activities are perhaps controlled by the local concentration of angiotensin or by ions. The "macula densa theory" of Thurau (1973, 1974, 1975) might have a bearing on the ionic control of APA in the JGA; this theory states that changing ion conditions in the distal tubule influence renin secretion via the macula densa (Thurau et al. 1972) and also influence ANG II formation in the JGA and thus glomerular blood flow ("tubulo-glomerular balance"). It is conceivable that APA activities are also controlled by calcium, sodium, or chloride ions that reach the JGA via the macula densa. This is evidenced by our findings on the dependence of this enzyme on calcium chloride and sodium chloride.

4.2.2 Renal Corpuscle

Only APA occurs in the renal corpuscle of the rat, while both APA and APM occur there in the mouse. The glomerulus contains only APA in both species. This agrees with other investigators, who were able to demonstrate APA (Glenner et al. 1962, Lojda and Gossrau 1980, Kugler 1981b) but not APM (Wachsmuth 1968, Wachsmuth and Torhorst 1974, Wachsmuth and Donner 1976, Kugler 1981b) in the glomerulus. Bowman's capsule in the rat consists of a monolayer squamous epithelium that is devoid of APA and APM. In mice, however, Bowman's capsule is formed by an isoprismatic epithelium with brush borders (tubule-like cells, Von Möllendorf 1930, Helmholz 1935, Hanker et al. 1975, Kugler 1981b); this is more prominent in males than females. These brush borders contain both APA and APM.

The occurrence of APA in the glomerulus has two possible functional implications, one relating to the N-terminal hydrolysis of filterable ANG I and II, the other to the regulation of ANG II effects on the glomerulus by ANG II degradation.

The quantity of intact peptide hormone that reaches the proximal tubule is regulated by the degradation of filterable angiotensin. Our ultracytochemical investigations point to the cell membranes of endothelial cells and podocytes as the sites of ANG I and II degradation, as APA is demonstrable mainly at these membranes. The localization of APA at cellular surfaces ensures that it will have intimate contact with filterable angiotensin. This aspect of glomerular angiotensin degradation appears to be essential, for ANG II in the S_1 proximal tubule segment is believed to be significant for sodium reabsorption (see below).

Because ANG II exerts an effect on the glomerulus itself, the degradation of angiotensin by APA in the glomerulus must also be considered from this standpoint. The function of ANG II in the glomerulus is to modulate both the contractile state of the glomerulus and glomerular filtration properties.

Scanning electron microscopic studies have shown that ANG II causes contraction of the glomerulus (Hornych et al. 1971, 1972). The glomerular capillaries (Constandinides and Robinson 1969, Spinelli et al. 1973) and the mesangial cells are presumably involved in this contraction. In vitro studies have confirmed that ANG II causes contraction in mesangial cells (Ausiello et al. 1979, 1980, Foidart et al. 1980). ANG II has been demonstrated autoradiographically in mesangial cells (Osborne et al. 1975). Our investigations have shown that APA is not clearly demonstrable in mesangial cells.

With regard to the effects of ANG II on glomerular filtration properties, it is known from physiologic studies that this peptide hormone lowers the glomerular permeability coefficient (Blantz et al. 1976, Baylis and Brenner 1978, Blantz 1980). A permeability change in the glomerulus under the influence of angiotensin is also expressed in ANG II-induced proteinuria (Eisenbach et al. 1975).

The changes produced by angiotensin in the glomerulus, which affect both its contractility and filtration properties, could be mediated by angiotensin receptors that have been demonstrated there (Sraer et al. 1974, Beaufils et al. 1976, Srauer et al. 1977). The APA demonstrated by us at cell membranes of the glomerulus counteracts these contractile and permeability effects by degrading ANG II. Our quantitative histochemical results on APA in subcapsular glomeruli of the rat and mouse must also be discussed from this standpoint. In both species we find a significant increase in APA activities after low-sodium feeding, and in the rat after adrenalectomy. This activity increase must be viewed in connection with the increased production of ANG II under these circumstances. It has been proved in previous studies that the production of renin, and thus of ANG II, is increased after adrenalectomy and sodium deprivation (Gross and Sulzer 1957, Schaechtelin et al. 1963, Gross et al. 1965, Peart 1965, Bareiss and Kracht 1969, Danda and Deveny 1971, Bucher and Kaissling 1973 lit., Churchill et al. 1973 lit., Nishimura 1980 lit., Nakane et al. 1980). In the presence of increased ANG II, the elevated APA activities could serve to prevent an interruption (due to contraction of glomerulus by ANG II) of glomerular blood flow and filtration.

The fact that APA activities in the glomerulus are also significantly elevated in both species after sodium loading is more difficult to explain, because the production of renin, and thus of ANG II, is greatly depressed under these conditions (Bareiss and Kracht 1969, Shade et al. 1972, Bucher and Kaissling 1973 lit., Kotchen et al. 1978, Nishimura 1980 lit., Nakane et al. 1980). However, this finding can be understood if we consider the results of physiologic studies showing that the perfusion of subcapsular glomeruli is increased in animals fed a high-sodium diet (Deetjen 1976 lit.). Our measurements were also performed in subcapsular glomeruli, and we presume that the increased rate of ANG II hydrolysis with a low blood angiotensin must greatly curtail the delivery of ANG II to glomerular receptors, so that glomerular contraction is reduced and maximum glomerular blood flow can occur.

On the whole, then, the glomerulus must be viewed not just as a receptor organ, but also as a filtration site for angiotensin.

An interesting finding in the podocytes of mice should also be mentioned. A reaction product referable to APA has been demonstrated in the nuclear membrane region of these cells. This raises the question of a possible intracellular function of ANG II. This question cannot be answered for the podocytes. However, ANG II has been demonstrated in the nuclei of smooth muscle cells (Robertson and Khairallah 1971), and increased nucleic acid synthesis in the myocardium has been induced by ANG II (Khairallah et al. 1972). If angiotensin performs a similar function in podocytes, then APA could indirectly regulate angiotensin-induced nucleic acid synthesis through its degradation of this hormone.

4.2.3 Proximal Tubule

Both APA and APM occur in the brush border of the proximal tubule in the rat and mouse (cf. Glenner et al. 1962, Lojda and Gossrau 1980, Kugler 1981b lit.).

According to quantitative histochemical measurements in male rats, APA activities are highest in the S_3 segment. Setting this equal to 100%, we find that the APA activity in the S_2 segment is about 63%, and in the S_1 segment, about 14% (based on equal reaction volumes and not on the actual total volumes of these tubule segments). A marked sex difference exists, for while APA activities are highest in the terminal portions of the S_3 segment in male rats, this segment is nearly negative for APA in female rats. In mice, the S_3 segments are APA-negative in both sexes.

APM is demonstrable in all three proximal tubule segments of the rat and mouse, the activities increasing from S_1 to S_3. Still, APM shows higher activities in the S_1 segment than APA. APM activities are lower in female rats than in males, especially in the S_3 segment.

The sex-dependent distribution of the two aminopeptidases in rats has also been observed with other peptidases and apparently bears a relation to the sex-dependent proteinuria in this species (Jedrzejewski and Kugler, to be published).

Angiotensin degradation in the proximal tubule is important in two respects — first because angiotensins are hydrolyzed mainly in this part of the nephron, and second because ANG II exerts a specific action there.

The two brush-border peptidases, APA and APM, appear to be essentially responsible for the ANG II degradation known from renal perfusion studies. According to these investigations, the angiotensins undergo a rapid hydrolysis in the kidney, and only small amounts are excreted intact in the urine (Pullman et al. 1975, Carone et al. 1979, 1980). It has been confirmed in isolated proximal tubule preparations that the brush borders represent the major sites of angiotensin degradation (Peterson and Carone 1974, Peterson et al. 1977, Pullman et al. 1978, Silbernagl 1978, Peterson et al. 1979, Carone et al. 1979, 1980). Although the high APA and APM activities in the S_3 segment lead to an increased liberation of amino acids, it is reported that amino acid absorption takes place mainly in the S_1/S_2 segment (Deetjen 1976, Silbernagl 1980, Silbernagl et al. 1980).

Because APA is thought to be relatively specific in its removal of N-terminal aspartic and glutamic acid (McDonald and Schwabe 1979), and APM hydrolyzes these acids very slowly (Kenny 1979 lit.) while hydrolyzing other amino acids rapidly and nonspecifically (Kenny 1979 lit.), it may be assumed that APA and APM hydrolyze angiotensin in a sequential manner: First APA removes aspartic acid from the angiotensin molecule, and then APM splits off subsequent amino acids in the remainder of the molecule.

A sequential hydrolysis of the angiotensin molecule by APA and APM is also indicated by our qualitative histochemical findings in rat kidneys with the substrate α-L-Asp-Arg-MNA. This substrate contains the first two amino acids of the ANG I and ANG II molecule. Attack of this substrate is histochemically demonstrable at the sites where APA and APM are jointly localized, specifically in brush borders of the proximal tubule. The glomerulus shows no reaction. This suggests that first APA removes the N-terminal aspartic acid, and then APM splits the arginine, resulting in the formation of azo dye at the site where MNA is liberated, i.e., in the brush borders. It is also conceivable that APM cleaves aspartic acid and arginine from the artificial substrate, thus giving rise to the same reaction pattern. However, this hypothesis is weakened by our finding that the attack of α-L-Asp-Arg-MNA is activated by $CaCl_2$. The calcium-activated APA removes increased amounts of aspartic acid in the presence of calcium, thereby increasing the Arg-MNA available for hydrolysis by APM. Because APA is the rate-limiting step in this reaction, greater activity is demonstrated under the influence of calcium.

Angiotensin degradation in the proximal tubule must be considered from yet another standpoint, however. This is the specific effect exerted by ANG II on the proximal tubule. According to Harris and Young (1977) and Harris (1979), ANG II promotes sodium reabsorption as a function of the concentration of the peptide hormone. Results in the literature are contradictory, however (cf. Thurau 1975 lit.). Angiotensin receptors demonstrable in the proximal tubule (Freedlender and Goodfriend 1977) could mediate the effect of ANG II on sodium reabsorption.

Assuming that ANG II does influence sodium uptake in the proximal tubule, the APA and APM reactions in the S_1 and S_2 segments would presumably play a key regulatory role, as these segments constitute the main sites of sodium reabsorption; the S_3 segment would represent a less specific site of angiotensin degradation. The low APA activities in the S_1 segment are of particular interest in this context, in that the resulting low hydrolysis of ANG II allows greater sodium uptake into the proximal tubule cell.

4.2.4 Renal Homogenate

According to our biochemical fluorometric measurements of renal homogenate in rats, APA and APM activities are decreased after adrenalectomy and high- and low-sodium feeding compared with control animals; the decreases in APA and APM after high-sodium feeding and in APA after adrenalectomy are statistically significant. Mice, by contrast, show no statistically significant change compared with controls, except in APM after high-sodium feeding. However, the generally lower specific APA and APM activity in the kidneys of mice is interpreted as a true species difference.

Data collected under similar experimental conditions in investigations of angiotensinases in renal homogenates agree with our results, in the respect that angiotensinase activities are decreased after sodium deprivation and sodium loading (Jelinek et al. 1971).

Ledingham and Leary (1974 lit.) state that studies of homogenate can provide no information on the degradation of angiotensins — that such information can be obtained only by perfusion studies of the organ in question. Because we are familiar with the histochemical localization of these enzymes (APA and APM), and thus know that they can come into direct contact with the blood (glomerulus) and primary urine (proximal tubule), this statement is refuted. This also means that histochemical studies in tissue sections and biochemical investigations in tissue homogenate can meaningfully supplement each other.

There is an obvious discrepancy between our quantitative histochemical data in subcapsular glomeruli and our biochemical fluorometric data in renal homogenate. This illustrates the limited value of homogenate studies in organs with a high degree of structural differentiation. Homogenization mixes the enzyme activities of the different parts of the nephron, thereby "averaging out" the peptidase activities over the whole kidney. Thus it can be determined only that the renal capacity for angiotensin degradation decreases in rats under the various experimental conditions, but remains largely unchanged in mice. Another problem is that the enzymes, especially APM, have functions other than angiotensin hydrolysis, thus making it still more difficult to interpret the results of renal homogenate studies.

4.3 Conclusion

In summary, our work has shown that the enzymes we investigated, APA and APM, are angiotensin-degrading aminopeptidases. Their occurrence in the kidney is limited mainly to the proximal part of the nephron. While APA is an aminopeptidase which removes L-glutamic and L-aspartic acids, APM possesses a broader activity spectrum. In terms of angiotensin degradation, this apparently means that APA mainly removes the N-terminal aspartic acid of ANG I and II, while APM mainly hydrolyzes (Des-Asp) angiotensin (e.g., ANG III) into amino acids. Thus, APA starts the breakdown of the angiotensin molecule from the N-terminal side (ANG I, II), while APM continues the process in sequential fashion.

The two aminopeptidases show a considerable difference in their localization in the nephron: APA is demonstrable in the glomerulus and JGA, while APM is not. Both peptidases occur in the brush border of the proximal tubule. Assuming that angiotensin degradation is accomplished by these two enzymes alone, it may be inferred that

ANG III is formed from ANG II in the glomerulus and JGA, while complete hydrolysis of angiotensins can occur in the proximal tubule. The functional importance of ANG III in the glomerulus or JGA, if any, remains unsettled. It is known that ANG III has less vasopressor activity than ANG II but, like ANG II, stimulates adrenal mineralocorticoid synthesis.

ANG II has different actions in the glomerulus, JGA, and proximal tubule. The angiotensin receptors demonstrable in the glomerulus and proximal tubule probably mediate its effects. It is assumed that APA and APM are located in close proximity to these receptors, for such proximity has been demonstrated between angiotensin receptors and angiotensinases in the rabbit aorta. Thus, APA in the glomerulus (and perhaps in the JGA) as well as APA and APM in the proximal tubule could regulate, through their degrading action, the amount of local ANG II presented to the receptors, or modify the duration of action of the angiotensin molecule according to their activities. At the same time, APA and perhaps APM could mark the site of angiotensin receptors in the nephron, assuming that angiotensinases always occur at the site of the angiotensin receptors. The membrane-bound occurrence of angiotensin receptors is consistent with the localization of APA, for APA is demonstrable mainly at cell membranes. It is also noteworthy that APA is a calcium-dependent enzyme, and that ANG II reportedly mediates calcium-ion shifts into the cells. A connection between the calcium influx produced by angiotensin and APA activities is speculative, however. Regarding the control of APA activities, it may be presumed that it is accomplished by small ions or by the supply of substrate in the form of angiotensins.

The occurrence of APA in the JGA is of special physiologic relevance. Besides renin and angiotensin, APA can now also be demonstrated in epithelioid cells. ANG II influences renin production, perhaps through a negative feedback mechanism; i.e., as the concentration of ANG II rises, renin production is inhibited. According to our investigations APA intervenes in this feedback process. Depending on APA activity, the local signal is modified by the breakdown of ANG II, which in turn alters renin production. Adrenalectomized animals may offer an example of such regulation. Here the APA activity in epithelioid cells is high, implying a rapid, local degradation of ANG II with no inhibition of renin production.

5 Summary

Qualitative and quantitative histochemical, ultracytochemical, and biochemical fluorometric methods were used to investigate aminopeptidase A (APA, E.C. 3.4.11.7) in the kidney of the rat and mouse and compare it with aminopeptidase M (APM, E.C. 3.4.11.2), which was subjected to qualitative histochemical and biochemical fluorometric study. Methodological studies were done to characterize the enzymes in terms of angiotensin degradation, and animal studies to explore the possible functional significance and regulation of these enzymes in the kidney.

The methodological studies of APA, which were done mainly in the kidneys of male rats and involved various tissue treatments and demonstration methods, showed that the following procedure is favourable for qualitative and quantitative histochemical demonstrations: Fixation of un-pretreated cryostat sections in 100% acetone at $-25\,°C$, followed by incubation of the sections at +30 °C in a reaction medium containing 1.5–1.65 mM α-L-Glu-MNA, 1.5 or 2.5 mM $CaCl_2$, 1.1 mM FBB (purest-grade), and tris-maleate buffer, pH 6.5 (pH of reaction medium = 7). Our histochemical and biochemical work (renal homogenate) showed that APA is a calcium-ion-dependent enzyme, while APM is not.

To clarify the functional importance of APA and APM in the kidney, their activities were measured under the influence of angiotensins. The quantitative histochemical measurements were done kinetically with an instrumental setup consisting of a microdensitometer and a computer-supported morphometric system (Kugler 1981a). We selected the glomerulus as the measuring site, since besides APA it contains no APM or other peptidases that could degrade angiotensins. Using the Lineweaver-Burk plot, we determined a K_m of 0.23 mM for the APA in rat glomeruli. ANG I, II, and III competitively inhibit APA in the rat and mouse glomeruli (K_i for ANG II in rat glomeruli = 0.14 mM).

For control and comparison purposes, biochemical fluorometric measurements were performed in microdissected rat glomeruli. The results were similar to the quantitative histochemical studies (K_m = 0.23 mM, competitive inhibition of APA by ANG I and II). Biochemical fluorometric measurements in renal homogenates (with 2-naphthylamide derivatives as substrates), which represent mixed-enzyme tissue preparations containing a variety of peptidases besides APA and APM, showed a K_m of 0.13 mM for APA and competitive inhibition by ANG II (K_i = 0.015 mM), and a K_m of 0.24 mM and competitive inhibition by ANG III (K_i = 0.003 mM). The remaining two angiotensins showed noncompetitive inhibition of APA (ANG I, III) and APM (ANG I, II) in this preparation.

With the variety of measuring techniques it was possible to show that APA is equivalent to angiotensinase A (splitting off the N-terminal aspartic acid from ANG I and II), and that both APA and APM participate in angiotensin degradation in the kidney, APA initiating the breakdown of ANG I and II, and APM possibly continuing it in sequential fashion.

Histochemically, APA is localized mainly in the glomeruli and brush borders of the proximal tubule, and APM in the brush borders of the proximal tubule. The JGA was identified as a previously unknown reaction site for APA in the kidney. The distribution of APA is different in the JGA of the rat and mouse. In the rat the Goormaghtigh's cell nests, especially of subcapsular glomeruli, react with varying activities, while in the mouse the epithelioid cells of the afferent arteriole do so. Ultracytochemical studies of glomeruli and JGA have shown that APA is localized mainly at the cell membranes of podocytes and endothelial cells (rat and mouse) and of epithelioid cells (mouse) and Goormaghtigh's cells (rat). In the epithelioid cells of mice, reaction product is also observed intracellularly in lysosomal structures. The two aminopeptidases show species and sex differences in the proximal tubule. While the terminal portions of the S_3 segments show the highest APA activities in male rats, they are nearly negative in females. Moreover, the brush-border reactions of APM are higher in the subcortical zone of male rats than in that of females. In mice, APM is demonstrable in the S_3 segment, but not APA.

Animal experiments were conducted with rats and mice (low- and high-sodium diet, adrenalectomy) to clarify the possible functional importance and regulation of APA and APM in the kidney. Quantitative histochemical measurements in subcapsular glomeruli showed a statistically significant rise in APA activities under all experimental conditions. By contrast, the activities of APA and APM in renal homogenate decreased in the rat, though not significantly, or remained unchanged in the mouse. The most pronounced APA changes were demonstrated histochemically in the JGA following adrenalectomy. In the rat the activities of the Goormaghtigh's cell nests increased, and the epithelioid cells, which were negative in normal animals, showed a positive reaction. In mice, high APA activities were demonstrated in hypertrophic epithelioid cell complexes of the afferent arteriole, as well as to some degree of the efferent arteriole.

In rats and mice fed a low-sodium diet, there was a slight but detectable activity increase in the APA-positive areas of the JGA. These areas showed an activity decrease after high-sodium feeding.

The findings of our localization studies and animal experiments suggest that APA and APM, by their degradation of angiotensin, are important in the regulation of intrarenal angiotensin effects. Of particular significance is the occurrence of APA in the JGA and glomerulus, where angiotensins modulate renin secretion and glomerular blood flow. ANG II induces contraction in the contractile components of the JGA (afferent and efferent arteriole, Goormaghtigh's cells) and the glomerulus (mesangium), thereby altering the glomerular blood flow; in the epithelioid cells it acts to inhibit renin production. However, these effects may not be mediated by ANG II itself, but by calcium ions. APA may counteract these effects by its degradation of this peptide hormone. Taking as an example adrenalectomy, which leads to an increased production of renin and thus of ANG II, this would mean that the heightened APA activities in the JGA and glomeruli act to prevent the inhibition of renin secretion and help to maintain the glomerular blood flow.

References

Allen E (1922) The oestrus cycle in the mouse. Am J Anat 30:297–371

Ausiello DA, Kreisberg JI, Roy C (1979) Contraction of cultured rat glomerular mesangial (MS) cells after stimulation with angiotensin II (AG II) and arginine vasopressin (AVP). Kidney Int 16:804

Ausiello DA, Kreisberg JI, Roy C, Karnovsky MJ (1980) Contraction of cultured rat glomerular mesangial cells after stimulation with angiotensin II and arginine vasopressin. J. Clin Invest 63:754–760

Bakhle YS (1974) Converting enzyme in vitro measurement and properties. In: Page JH, Bumpus FM (eds) Angiotensin. Springer, Berlin Heidelberg New York pp 41–80

Bareiss W, Kracht J (1969) Beziehung zwischen juxtaglomerulärem Apparat und Zona glomerulosa unter NaCl-Belastung, -Mangel und Durst. Endokrinologie 54:327–343

Bargmann W (1978) Niere und ableitende Harnwege. Handbuch der mikroskopischen Anatomie des Menschen, vol 7/5. Springer, Berlin Heidelberg New York

Baylis C, Brenner BM (1978) Modulation by prostaglandin synthesis inhibitors of the actions of exogenous angiotensin II on glomerular ultrafiltration in the rat. Circ Res 43:889–898

Beaufils M, Sraer J, Lepreux C, Ardaillon R (1976) Angiotensin II binding to renal glomeruli from sodium-loaded and sodium depleted rats. Am J Physiol 230:1187–1193

Bisswanger H (1979) Theorie und Methoden der Enzymkinetik. Verlag Chemie, Weinheim Deerfield Beach, FL Basel

Blantz RC (1980) The glomerulus, passive filter or regulatory organ? Klin Wochenschr 58:957–964

Blantz RC, Konnen KS, Tucker BJ (1976) Angiotensin II effects upon glomerular microcirculation and ultrafiltration coefficient of the rat. J Clin Invest 57:419–434

Bohr DF (1974) Angiotensin on vascular smooth muscle. In: Page IH, Bumpus FM (eds) Angiotensin. Springer, Berlin Heidelberg New York pp 424–440

Britton KE (1968) Renin and renal autoregulation. Lancet 2:329–334

Bucher O, Kaissling B (1973) Morphologie des juxtaglomerulären Apparates. Verh Anat Ges 67:109–136

Bucher O, Reale E, Erkocak A (1967) Die epitheloiden Zellen des juxtaglomerulären Apparates der Meerschweinchen-Niere vor und nach bilateraler Adrenalektomie. Anat Anz 120:243–249

Carone FA, Peterson DR, Oparil S, Pullman TN (1979) Renal tubular transport and catabolism of proteins and peptides. Kidney Int 16:271–278

Carone FA, Peterson DR, Oparil S, Pullman TN (1980) Renal tubular transport and catabolism of small peptides. In: Maunsbach AB, Olsen TS, Christensen EI (eds) Functional ultrastructure of the kidney. Academic Press, London New York Toronto Sydney San Francisco pp 327–340

Celio MR, Ingami T (1981) Angiotensin II immunoreactivity coexists with renin in the juxtaglomerular granular cells of the kidney. Proc Natl Acad Sci USA 78:3897–3900

Churchill PC, Churchill MC, McDonald JK, Koskins A (1973) Renin release in anesthetized rats. Kidney Int. 4:273–279

Constantinides P, Robinson M (1969) Ultrastructural injury of arterial endothelium: II Effects of vasoactive amines. Arch Pathol 88:106–112

Cook WF (1971) Cellular localization of renin. In: Fisher JW (ed) Kidney hormones. Academic Press, New York London pp 117–128

Crabtree C (1940) Sex differences in structure of Bowman's capsule (of kidney) in mouse. Science 91:299

Dahlheim M, Schmidmeier E (1975) Determination and characterisation of angiotensinase activity of microdissected JGAs. Proc. 5th Int Congr Nephrol 1972. S Karger, Basel

Danda G, Deveny I (1971) Bilateral adrenalectomy and renin-angiotensin system. Acta Physiol Acad Sci Hung 39:335–341

Davidoff M, Schiebler TH (1981) Immunocytochemical demonstration of renin in the mouse and rat kidney after adrenalectomy. Histochemistry 72:453–457

79

Deetjen P (1976) Nierenphysiologie. In: Gauer OH, Kramer K, Jung R (eds) Physiologie des Menschen, vol 7. Urban und Schwarzenberg, Munich Berlin Wien, pp 1–102

Delange RJ, Smith EL (1971) Leucineaminopeptidase and other N-terminal exopeptidases. In: Boyer PD (ed) The enzymes, vol III, 3rd ed. Academic Press, New York London, pp 81–118

Dengler H, Reichel G (1960) Untersuchungen zur intrazellulären Lokalisation der Renin- und Hypertensinase-Aktivität. Experientia 16:36-38

Doyle AE, Louis WJ, Osborn EC (1967) Plasma angiotensinase activity on angiotensin II and analogues. Aust J Exp Biol Med Sci 45:41–50

Eisenbach GM, Van Liew JB, Boylan JW, Manz N, Huir P (1975) Effect of angiotensin on the filtration of protein in the rat kidney: a micropuncture study. Kidney Int 8:80–87

Faarup P (1971) Morphological aspects of the renin-angiotensin system. Acta Pathol Microbiol Scand [Suppl] 222:1–96

Finkielman S, Goldstein DJ, Fischer-Ferraro C, Diaz A, Nahmod VE (1972) In vitro production of angiotensin and renin release by isolated glomeruli. Medicina (Buenos Aires) 32 [Suppl]: 37–39

Foidart JB, Dechenne C, Dubois CH, Mahien PR (1980) Tissue culture of rat glomerular endothelial cells. Kidney Int 18:136

Freedlender AE, Goodfriend TL (1977) Angiotensin receptors and sodium transport in renal tubules. Fed Proc 36:481

George SG, Kenny AJ (1973) Studies on the enzymology of purified preparations of brush broder from rabbit kidney. Biochem J 134:43–57

Glenner GG, Folk JE (1961) Glutamylpeptidases in rat and guinea-pig kidney slices. Nature 192: 338–340

Glenner GG, McMillan PMY, Folk LF (1962) A mammalian peptidase specific for the hydrolysis of N-terminal α-L Glutamyl and aspartyl residues. Nature 194:867

Goormaghtigh N (1932) Les segments nervo-myo-artériels juxta-gloméruliares du rein. Arch Biol (Liege) 43:575–591

Gorgas K (1978) Struktur und Innervation des juxtaglomerulären Apparates der Ratte. Adv Anat Embryol Cell Biol 54:2

Granger P, Dahlheim H, Thurau K (1969) Renin-, Angiotensinase- und Converting-Enzym-Bestimmungen an einzelnen mikrodisserzierten juxtaglomerulären Apparaten und Teilen des Nephrons. Pfluegers Arch 312:R87

Granger P, Dahlheim H, Thurau K (1972) Enzyme activities of single juxtaglomerular apparatuses in the rat kidney. Kidney Int 1:78–88

Gross F, Sulzer F (1957) Der Einfluß der Nebenniere auf die blutdrucksteigernde Wirkung von Renin und auf pressorische Substanzen in den Nieren. Arch Exp Pathol Pharmakol 230:274–283

Gross F, Brunner M, Ziegler M (1965) Renin-Angiotensin system, aldosterone and sodium balance. Recent Prog Horm Res 21:119–167

Hanker JS, Preece JW, Mac Rae EK (1975) Cytochemical correlates of structural sexual dimorphism in glandular tissues of the mouse. I. Studies of the renal glomerular capsule. Histochemistry 44:225–244

Hanson M, Hütter M-J, Mannsfeldt M-G, Kretschmer K, Sohr Ch (1967) Zur Darstellung und Substratspezifität einer von der Leucinaminopeptidase unterscheidbaren Aminopeptidase aus Nierenpartikeln. Hoppe Seylers Z Physiol Chem 348:680–688

Harris PJ (1979) Stimulation of proximal tubular sodium reabsorption by Ile^5-Angiotensin II in the rat kidney. Pfluegers Arch 381:83–85

Harris PJ, Young JA (1977) Dose-dependent stimulation and inhibition of proximal tubular sodium reabsorption by angiotensin II in the rat kidney. Pfluegers Arch 367:295–297

Helmholz H (1935) The presence of tubular epithelium within the glomerular capsule in mammals. Proc Staff Meet Mayo Clin 10:110

Hess R (1965) Arylamidase activity related to angiotensinase. Biochim Biophys Acta 99:316–324

Hierholzer K (1977) Das Renin-Angiotensin-Aldosteronsystem. In: Gauer OH, Kramer K, Jung R (eds) Physiologie des Menschen, vol 20. Urban und Schwarzenberg. Munich Wien Baltimore, pp 219–261

Hodge RL, NG ICKF, Vane JR (1967) Disappearance of angiotensin from the circulation of the dog. Nature 215:138–141

Holt SJ (1959) Factors governing the validity of staining methods for enzymes, and their bearing upon the Gomori acid phosphatase technique. Exp Cell Res [Suppl] 7:1–27

Horky K, Rojo-Ortega JM, Rodriguez J, Boucher R, Genest J (1971) Renin, renin substrate and angiotensin I-converting enzyme in the lymph of rats. Am J Physiol 220:307–311

Hornych H, Beaufils M, Richet G (1971) Effects de l'angiotensine exogène sur les capillaires des glomérules corticeux et juxtamédullaires du Rat: étude en microscopic électronique à balayage. C R Acad Sci [D] (Paris) 273:1129–1131

Hornych H, Beaufils M, Richet G (1972) The effect of exogenous angiotensin on superficial and deep glomeruli in the rat kidney. Kidney Int 2:336–343

Jedrzejewski K, Kugler P (to be published) Zur Histochemie und Biochemie von aminosäure- und peptidabspaltenden Enzymen in der Rattenniere. Verh Anat Ges

Jelinek J, Mannel H, Gross F (1971) Inactivation of angiotensin II by rat kidney homogenates. Arch Pharm (Weinheim) 268:446–457

Johnson JA, Anderson RR (1980) The renin-angiotensin system. Plenum Press, New York London

Johnson DC, Ryan JW (1968) Degradation of angiotensin II by a carboxypeptidase of rabbit liver. Biochim Biophys Acta 160:196–203

Kalsner S, Nickerson M (1968) A method for the study of mechanisms of drug disposition in smooth muscle. Can J Physiol Pharmacol 46:719–730

Kenny AJ (1979) Proteinases associated with cell membranes. In: Barrett AJ (ed) Proteinases in mammalian cells and tissues. North-Holland, Amsterdam Oxford New York, pp 393–444

Khairallah PA, Hall MM (1977) Angiotensinase. In: Genest JG, Koin E, Kuchel O (eds) Hypertension. McGraw Hill, New York, pp 179–183

Khairallah PA, Page JH (1967) Plasma angiotensinases. Biochem Med 1:1–8

Khairallah PA, Bumpus FM, Page JH, Shemby RR (1963) Angiotensinase with a high degree of specificity in plasma and red cells. Science 140:672–674

Khairallah PA, Robertson AL, Davila D (1972) Effects of angiotensin II on DNA, RNA, and protein synthesis. In: Genest J, Koiw E (eds) Hypertension '72. Springer Berlin Heidelberg New York, pp 218–220

Klaus D, Kaffarnick H, Pfeil H (1963) Untersuchungen über die Serum-Angiotensinase II. Klinische Ergebnisse beim Hochdruck und bei Lebererkrankungen. Klin Wochenschr 41:380–388

Kotchen TA, Galla JH, Luke RG (1978) Contribution of chloride to the inhibition of plasma renin by sodium chloride in the rat. Kidney Int 13:201–207

Kriz W, Taugner R (1977) Speculations on the functional significance of the Goormaghtigh cells. Kidney Int 11:217

Kroon DB (1960) Origin of the PAS-positive granulated ε-cells of the juxtaglomerular apparatus. Acta Anat (Basel) 41:138–156

Kugler P (1981a) Kinetic and morphometric measurements of enzyme reactions in tissue sections with a new instrumental setup. Histochemistry 71:433–449

Kugler P (1981b) Localization of aminopeptidase A (angiotensinase A) in the rat and mouse kidney. Histochemistry 72:269–278

Kurtz AB, Wachsmuth ED (1969) Identification of plasma angiotensinase as aminopeptidase. Nature 221:92–93

Leary WP, Ledingham JG (1969) Removal of angiotensin by isolated perfused organs of the rat. Nature 222:959–960

Leckie B, Gavras H, McGregor J, McElwee G (1972) The conversion of angiotensin I to angiotensin II by rabbit glomeruli. J Endocrinol 55:229–230

Ledingham JG, Leary WP (1974) Catabolism of angiotensin II. In: Page JH, Bumpus FM (eds) Angiotensin. Springer, Berlin Heidelberg New York, pp 111–125

Lee JC, Hurley S, Hopper J (1966) Secretory actitivy of the juxtaglomerular granular cells of the mouse: morphologic and enzyme histochemical observations. Lab Invest 15.1459–1476

Lehky P, Lisowski J, Wolf DP, Wacker H, Stein EA (1973) Pig kidney particulate aminopeptidase a zinc metalloenzyme. Biochim Biophys Acta 321:274–281

Lever AF, Peart WS (1962) Renin and angiotensin-like activity in renal lymph. J Physiol (Lond) 160:548–563

Lineweaver H, Burk D (1934) The determination of enzyme dissociation constants. J Am Chem Soc 56:658–666

Lojda Z, Gossrau R (1980) Study on aminopeptidase A. Histochemistry 67:267−290

Lojda Z, Gossrau R, Schiebler TH (1979) Enzyme histochemistry. A laboratory manual. Springer, Berlin Heidelberg New York

Matsunaga M (1971) Nature of lysosomal angiotensinase activity. Jpn Circ J 35:333−338

Matsunaga M, Masson GMC (1970) Hepatic and renal neutral angiotensinases. Experientia 26:1297−1298

Matsunaga M, Kira J, Saito N, Ogino K, Takayasu M (1968) Acid angiotensinase, renin and adenosinetriphosphatase in rat kidney lysosomes. Jpn Circ J 32:137−143

Matsunaga M, Saito N, Kira J, Ogino K, Takayasu M (1969) Acid angiotensinase as a lysosomal enzyme. Jpn Circ J 33:545−551

Maunsbach AB (1966) Observations on the segmentation of the proximal tubule in the kidney. Comparison of results from phase contrast, fluorescence and electron microscopy. J Ultrastruct Res 16:239−258

McDonald JK, Schwabe C (1979) Intracellular exopeptidases. In: Barrett AJ (ed) Proteinases in mammalian cells and tissues, North-Holland, Amsterdam Oxford New York, pp 311−391

McGadey J (1970) A tetrazolium method for nonspecific alkaline phosphatase. Histochemie 23:180−184

McGiff JC (1968) Tissue hormones: Angiotensin, bradykinin and the regulation of regional blood flows. Med Clin North Am 52:263−281

Nagatsu IT, Gillespie IL, George JM, Folk JE, Glenner GG (1965) Serum aminopeptidase "angiotensinase" and hypertension. II. Aminoacid B-naphthylamid hydrolysis by normal and hypertensive serum. Biochem Pharmacol 14:853−861

Nagatsu IT, Yamamoto T, Glenner GG, Mehl JW (1970) Purification of aminopeptidase A in human serum and degradation of angiotensin II by the purified enzyme. Biochim Biophys Acta 198:255−270

Nakane H, Nakane Y, Corvol P, Menard J (1980) Sodium balance and renin regulation in rats: role of intrinsic renal mechanisms. Kidney Int 17:607−614

Ng ICKF, Vane JR (1967) Conversion of angiotensin I to angiotensin II. Nature 216:762−766

Nishimura H (1980) Comparative endocrinology of renin and angiotensin. In: Johnson JA, Anderson RR (eds) The renin-angiotensin system. Plenum Press, New York London, pp 29−77

Osborne MJ, D'Auriac GA, Meyer P, Werzel M (1970) Mechanism of extraction of angiotensin II in coronary and renal circulations. Life Sci 9:859−867

Osborne MJ, Droz B, Meyer P, Morel F (1975) Angiotensin II: renal localization in glomerular mesangial cells by autoradiography. Kidney Int 8:245−254

Padykula HA, Herman E (1955) Factors affecting the activity of adenosine triphosphatase and other phosphatases as measured by histochemical techniques. J Histochem Cytochem 3:161−169

Page IH, Bumpus FM (1974) Angiotensin. Springer, Berlin Heidelberg New York

Page IH, McSwain B, Knapp GM, Andres WD (1941) The origin of reninactivator. Am J Physiol 135:214−222

Peach MJ, Levens NR (1980) Molecular approaches to the study of angiotensin receptors. In: Johnson JA, Anderson RR (eds) The renin-angiotensin system. Plenum Press, New York London, pp 171−194

Peart WS (1965) The renin-angiotensin system. Pharmacol Rev 17:143−182

Peterson DR, Carone FA (1974) Angiotensin II handling by isolated perfused proximal straight kidney tubules. Kidney Int 6:84 A

Peterson DR, Oparil S, Flouret G, Carone FA (1977) Handling of angiotensin II and oxytocin by renal tubular segments perfused in vitro. Am J Physiol 232:F319−F324

Peterson DR, Chrabaszcz G, Peterson WR, Oparil S (1979) Mechanism of renal tubular handling of angiotensin. Am J Physiol 236:F365−F372

Pfleiderer G, Celliers PG (1963) Isolierung einer Aminopeptidase aus Nierenpartikeln. Biochem Z 339:186−189

Pfleiderer G, Celliers PG, Stannlovic M, Wachsmuth ED, Determann H, Braunitzer G (1964) Eigenschaften und analytische Anwendung der Aminopeptidase aus Nierenpartikeln. Biochem Z 340:552−564

Plentl AA, Page IH, Davis WW (1943) The nature of renin activator. J Biol Chem 147:143−153

Pullman TN, Carone FA, Oparil S, Nakamura S (1978) Effects of constituent amino acids on tubular handling of microinfused angiotensin II. Am J Physiol 234:F325—F331

Pullman TN, Oparil S, Carone FA (1975) Fate of labeled angiotensin II microinfused into individual nephrons in the rat. Am J Physiol 229:747—751

Regoli D, Riniker B, Brunner HR (1963) The enzymatic degradation of various angiotensin II derivatives by serum, plasma or kidney homogenates. Biochem Pharmacol 12:637—646

Robertson AL, Khairallah PA (1971) Angiotensin II: rapid localization in nuclei of smooth and cardiac muscle. Science 172:1138—1139

Rouffignac C de, Bonvalet JP, Menard J (1974) Renin content in superficial and deep glomeruli of normal and salt-loaded rats. Am J Physiol 226:150—154

Rouiller C. Orci L (1971) The structure of the juxtaglomerular complex. In: Rouiller C, Muller A (eds) The kidney, vol 4. Academic Press, New York, pp 1—80

Ryan JW (1974) The fate of Angiotensin II. In: Page IH, Bumpus FM (eds) Angiotensin. Springer, Berlin Heidelberg New York, pp 81:104

Saito N, Sanada K, Matsunaga M, Mukaino S, Kira J, Ogino K, Takayasu M (1969) Amino acid analysis after degradation of angiotensin II by rat kidney lysosomes. Jpn Circ J 33:87—93

Schaechtelin G, Resoli D, Gross F (1963) Bio-assay of circulating renin-like pressor material by isovolumic cross circulation. Am J Physiol 205:303—306

Scherberich JE, Falkenberg F, Mondorf AW, Müller H, Pfleiderer G (1974) Biochemical and immunological studies on isolated brush border membranes of human kidney cortex and their membrane surface proteins. Clin Chim Acta 55:179

Schmid HE (1962) Renin, a physiologic regulator of renal hemodynamics? Circ Res 11:185—190

Silbernagl S (1978) The role of brush border peptidases in degradation and reabsorption of γ-glutamyl peptides and peptide hormones studies with microperfusion of the proximal tubule of rat kidney. Kidney Int 13:532

Silbernagl S (1980) Renal transport of amino acids. Kidney Int 17:399

Silbernagl S, Joost J, Jarosch E, Völkl H (1980) Characteristics of renal tubular reabsorption of "acidic" amino acids. Nieren-Hochdruckkr 5:205—206

Skeggs LT, Leutz KE, Kahn JR, Hochstrasser H (1968) Kinetics of the reaction of renin with nine synthetic peptide substrates. J Exp Med 128:13—34

Shade RE, Davis JO, Johnson JA, Witty RT (1972) Effects of renal arterial infusion of sodium and potassium on renin secretion in the dog. Circ Res: 31:719—727

Sokabe H (1974) Phylogeny of the renal effects of angiotensin. Kidney Int 6:263—271

Spinelli F, Torhorst J, Wirz H, Pehling G, Baum H (1973) Contractility of glomerular capillaries: Angiotensin II-mediated contraction. Kidney Int 3:273

Sraer JD, Sraer J, Ardaillon R, Mimonne O (1974) Evidence for renal glomerular receptors for angiotensin II. Kidney Int 6:241—246

Srauer J, Band L, Cosyns J-P, Veroust P, Nivez M-P, Ardaillon R (1977) High affinity binding of [125]J-angiotensin II to rat glomerular membranes. J Clin Invest 59:69—81

Stowe NR, Schnermann J, Hermle M (1979) Feedback regulation of nephron filtration rate during pharmacologic interference with the renin-angiotensin and adrenergic systems in rats. Kidney Int 15:473—486

Sutherland LE (1970) A fluorescent antibody study of juxtaglomerular cells using the freeze-substitution technique. Nephron 7:512—523

Taugner C, Poulsen K, Hackenthal E, Taugner R (1979) Immunocytochemical localization of renin in mouse kidney. Histochemistry 62:19—27

Taugner R, Hackenthal E, Rix E, Nobiling R, Poulson K (to be published) Immuncytochemical investigations on the renin-angiotensin-system: renin, angiotensinogen, angiotensin I, angiotensin II and converting enzyme in the kidney of mice, rats and tupaia. Kidney Int

Thurau K (1964) Renal hemodynamics. Am J Med 36:698—719

Thurau K (1973) Die Bedeutung des juxtaglomerulären Apparats für die Funktion des Nephrons. Verh Anat Ges 67:137—150

Thurau K (1974) Intrarenal action of angiotensin. In: Page IH, Bumpus FM (eds) Angiotensin. Springer, Berlin Heidelberg New York, pp 475—489

Thurau K (1975) Modification of angiotensin-mediated tubulo-glomerular feedback by extracellular volume. Kidney Int 8:S202—S207

Thurau K, Dahlheim H, Granger P (1970) On the local formation of angiotensin at the site of the juxtaglomerular apparatus. In: Heptinstall RH (ed) Proc intern congr nephrol 1969, Vol 1. S Krager, Basel New York, pp 24–30

Thurau K, Dahlheim H, Grüner A, Mason J, Granger P (1972) Activation of renin in the single juxtaglomerular apparatus by sodium chloride in the tubular fluid at the macula densa. Cir Res 30–31: II–182–186

Vecsei P, Hackenthal E, Ganten D (1978) The renin-angiotensin-aldosterone system. Klin Wochenschr [Suppl] 56:5–21

Von Deimling OH (1964) Die Darstellung phosphatfreisetzender Enzyme mittels Schwermetall-Simultan-Methode. Histochemie 4:48–55

Von Möllendorf W (1930) Der Exkretionsapparat. In: Handbuch der mikroskopischen Anatomie des Menschen, vol VII/1. Springer, Berlin

Wachsmuth ED (1967) Untersuchungen zur Struktur der Aminopeptidase aus Partikeln von Schweinenieren. Biochem Z 346:467–473

Wachsmuth ED (1968) Lokalisation von Aminopeptidase in Gewebeschnitten mit einer neuen Immunfluoreszenztechnik. Histochemie 14:282–296

Wachsmuth ED (1980) Assessment of immunocytochemical techniques with particular reference to the mixed aggregation immunocytochemical technique. In: Trends in enzyme histochemistry and cytochemistry. Ciba Foundation symposium 73. Excerpta Medica, Amsterdam Oxford New York, pp 135–160

Wachsmuth ED, Donner P (1967) Conclusions about aminopeptidase in tissue sections from studies of amino acid naphthylamide hydrolysis. Histochemistry 47:271–283

Wachsmuth ED, Torhorst A (1974) Possible precursors of aminopeptidase and alkaline phosphatase in the proximal tubules of kidney and the crypts of small intestine of mice. Histochemistry 38:43–56

Ward PE, Erdös EG (1977) Metabolism of kinins and angiotensins in the kidney. In: Haberland GL, Rohen JW, Suzuki T (eds) Kininogenases 4. FK Schattauer, Stuttgart New York, pp 107–110

Ward PE, Gedney CD, Dowben RM, Erdös EG (1975) Isolation of membrane-bound renal kallikrein and kininase. Biochem J 151:755–758

Ward PE, Erdös EG, Gedney CD, Dowben RM, Reynolds RC (1976) Isolation of membrane-bound renal enzymes that metabolize kinins and angiotensins. Biochem J 157:643–650

Werning C (1972) Das Renin-Angiotensin-Aldosteron-System. Thieme, Stuttgart

Winckler J (1970a) Zum Einfrieren von Gewebe in stickstoff-gekühltem Propan. Histochemie 23:44–50

Winckler J (1970b) Kontrollierte Gefriertrocknung von Kryostatschnitten. Histochemie 22:234–240

Yang HYT, Erdös EG, Chiang TS (1968) New enzymatic route for the inactivation of angiotensin. Nature 218:1224–1225

Zimmermann (1933) Über den Bau des Glomerulus der Säugerniere. Z Mikrosk Anat Forsch 32:176–278

Subject Index

Advances in Anatomy Embryology and Cell Biology

Editors: F. Beck, W. Hild, J. van Limborgh, R. Ortmann, J. E. Pauly, T. H. Schiebler

A Selection

Volume 67
H. Wolburg
Axonal Transport, Degeneration, and Regeneration in the Visual System of the Goldfish
1981. 28 figures. IX, 94 pages.
ISBN 3-540-10336-8

Volume 68
A. A. M. Gribnau, L. G. M. Geijsberts
Developmental Stages in the Rhesus Monkey (Macaca mulatta)
1981. 27 figures. VI, 84 pages.
ISBN 3-540-10469-0

Volume 69
L. Záborszky
Afferent Connections of the Medial Basal Hypothalamus
1982. 31 figures. VIII, 107 pages.
ISBN 3-540-11076-3

Volume 70
W. Pfaller
Structure Function Correlation on Rat Kidney
Quantitative Correlation of Structure and Function in the Normal and Injured Rat Kidney
1982. 23 figures. VIII, 106 pages.
ISBN 3-540-11074-7

Volume 71
L. Thuneberg
Interstitial Cells of Cajal: Intestinal Pacemaker Cells?
1982. 94 figures. Approx. 120 pages.
ISBN 3-540-11261-8

Volume 72
H. Breuker
Seasonal Spermatogenesis in the Mute Swan (Cygnus olor)
1982. 30 figures. 104 pages.
ISBN 3-540-11326-6

Volume 73
G. Zweers
The Feeding System of the Pigeon (Columba livia L.)
1982. 45 figures. Approx. 90 pages.
ISBN 3-540-11332-0

Volume 74
J. Altman, S. A. Bayer
Development of the Cranial Nerve Ganglia and Related Nuclei in the Rat
1982. 64 figures. Approx. 120 pages.
ISBN 3-540-11337-1

Volume 75
V. Grouls, B. Helpap
The Development of the Red Pulp in the Spleen
1982. 37 figures.
Approx. 70 pages.
ISBN 3-540-11408-4

Springer-Verlag
Berlin
Heidelberg
New York

Angiotensin

Editors: I. H. Page, F. M. Bumpus
With contributions by numerous experts.

1974. 70 figures. XIX, 591 pages.
(Handbook of Experimental Pharmacology, Volume 37)
ISBN 3-540-06276-9

Contents: Biological Production. - The Fate of Angiotensin. - Converting Enzyme in vitro Measurement and Properties. - Catabolism of Angiotensin II. - Structure-Activity Relationship in Angiotensin II Analogs. - Antagonists of Angiotensin II. - Tachyphylaxis to Angiotensin Immunogenicity and Antigenicity - Measurement of Renin and of Angiotensin. - Angiotensin Immunoassay. - Bioassay. - Plasma of Serum Vasopressor Peptides Other than Angiotensins. - Primary Aldosteronism. - Secondary Hyperaldosteronism. - Intermediary Metabolism. - The Renin-Angiotensin System in the Control of Aldosterone Secretion. - Aldosterone Regulation in Sodium Deficiency. Effects of Aldosterone on Blood Pressure, Water, and Electrolytes. - Adrenal Medulla. - Central Neurogenic Effects. - Peripheral Effects. - Angiotensin on Vascular Smooth Muscle. - Circulatory Effects. - Effects on Renal Circulation. - Intrarenal Action. - Morphological Effects on Arteries, on the Permeabilit of the Vascular Wall. - Biochemical Effects. - Some Possible Functions.

The Renin Angiotensin System in the Brain

A Model for the Synthesis of Peptides in the Brain

Editors: D. Ganten, M. Printz, M. I. Phillips, B. A. Schölkens

1982. 108 figures, 46 tables. XVII, 385 pages.
ISBN 3-540-11344-4

Contents: Renin and Converting Enzyme. - Angiotensinogen, Angiotensin, Angiotensin Receptors. - Central Effects of Angio tensin. - Nomenclature of the Renin-Angiotensin System. - Nomenclature of Experimental Hypertension. - Subject Index. Author Index.

The existence of an endogenous brain renin angiotensin system has been a subject of controversy and intensive research over th past ten years. Definite proof of the existence of the component of this enzyme peptide system is provided for the first time in th monograph. Leading researchers present the state of the art on the brain renin angiotensin system based on biochemical, pharmacological, physiological and endocrinological studies. The significance of these studies lies in their contribution to our understanding of the synthesis, distribution, physiology, and function of neuropeptides. Divergent functions can now be attri buted to the brain renin angiotensin system - ranging from its pathophysiologic role in hypertension to its modification on thi and memory - and making the information contained in this volume of profound interest to both basic scientists as well as to clinicians in internal medicine and psychiatry.

Springer-Verlag
Berlin
Heidelberg
New York